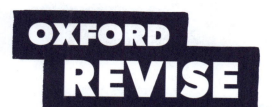

Oxford Revise

EDEXCEL A LEVEL

GEOGRAPHY

COMPLETE REVISION AND PRACTICE

David Alcock

Kyle McFarlane

Rebecca Priest

Paul Schofield

Lucy Scovell

Nadine Tunstall

OXFORD
UNIVERSITY PRESS

CONTENTS

 Shade in each level of the circle as you feel more confident and ready for your exam.

KEY

⚙ **Knowledge**

⇄ **Retrieval**

✎ **Practice**

CONTENTS

CONTENTS

HOW TO USE THIS BOOK

This book uses a three-step approach to revision: **Knowledge**, **Retrieval**, and **Practice**.
It is important that you do all three; they work together to make your revision effective.

 Knowledge

Knowledge comes first. Each topic is divided into
Knowledge Organisers. These are clear,
easy-to-understand, concise summaries of the content
that you need to know for your exam.

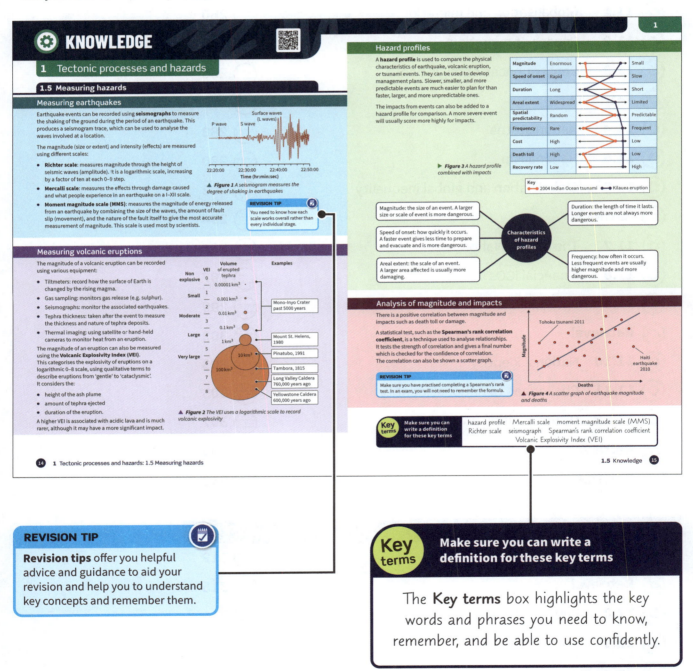

REVISION TIP

Revision tips offer you helpful
advice and guidance to aid your
revision and help you to understand
key concepts and remember them.

Key terms — Make sure you can write a
definition for these key terms

The **Key terms** box highlights the key
words and phrases you need to know,
remember, and be able to use confidently.

Retrieval

The **Retrieval questions** help you learn and quickly recall the information you've acquired. These are short questions and answers about the content in the Knowledge Organisers you have just revised. Cover up the answers with some paper and write down as many answers as you can from memory. Check back to the Knowledge Organisers for any you got wrong, then cover the answers and attempt all the questions again until you can answer *all* the questions correctly.

Make sure you revisit the Retrieval questions on different days to help them stick in your memory. You need to write down the answers each time, or say them out loud, for your revision to be effective.

Previous questions

Each Retrieval page also has some **Retrieval questions** from **previous sections**. Answer these to see if you can remember the content from the earlier sections. If you get the answers wrong, go back and do the Retrieval questions for the earlier sections again.

Practice

Once you are confident with the Knowledge Organisers and Retrieval questions, you can move on to the final stage: **Practice**.

Each chapter has **exam-style questions** to help you apply all the knowledge you have learned.

EXAM TIP

Exam tips show you how to interpret the questions, provide guidance on how to answer them, and give advice on how to secure as many marks as possible. Guidance is also offered on how to approach different command words.

Answers and Glossary

You can scan the QR codes at any time to access sample answers and mark schemes for the exam-style questions, a glossary containing definitions of the key terms, as well as further revision support, or go to go.oup.com/OR/Alevel/Ed/Geog

1 Tectonic processes and hazards

1.1 Plate tectonics

Hazard patterns and plate tectonics

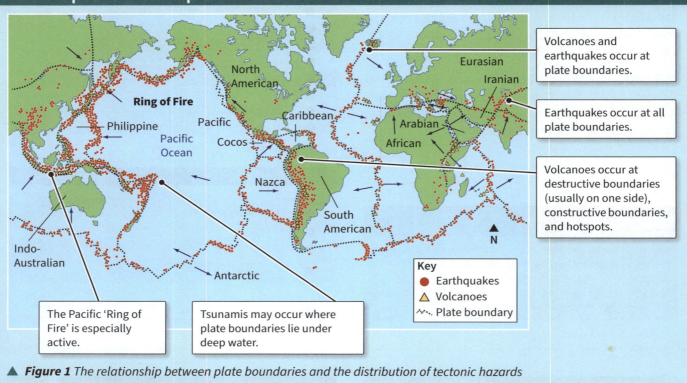

Volcanoes and earthquakes occur at plate boundaries.

Earthquakes occur at all plate boundaries.

Volcanoes occur at destructive boundaries (usually on one side), constructive boundaries, and hotspots.

The Pacific 'Ring of Fire' is especially active.

Tsunamis may occur where plate boundaries lie under deep water.

Key
- Earthquakes
- Volcanoes
- Plate boundary

▲ **Figure 1** *The relationship between plate boundaries and the distribution of tectonic hazards*

Earth's internal structure

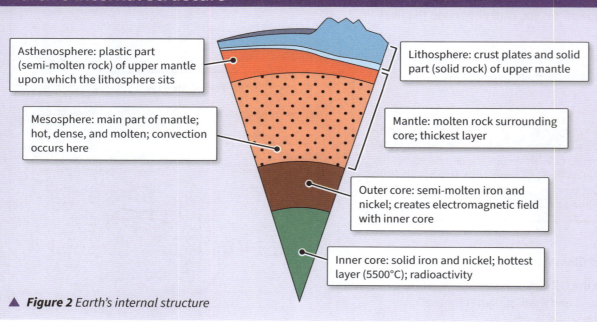

Asthenosphere: plastic part (semi-molten rock) of upper mantle upon which the lithosphere sits

Lithosphere: crust plates and solid part (solid rock) of upper mantle

Mesosphere: main part of mantle; hot, dense, and molten; convection occurs here

Mantle: molten rock surrounding core; thickest layer

Outer core: semi-molten iron and nickel; creates electromagnetic field with inner core

Inner core: solid iron and nickel; hottest layer (5500°C); radioactivity

▲ **Figure 2** *Earth's internal structure*

Plate tectonic theory

Plate tectonic theory combines the 1912 theory of **continental drift**, evidence that the sea floor is spreading at mid-ocean ridges, and **palaeomagnetism**.

- The Earth's crust (the **lithosphere**) is divided up into plates of different sizes (seven large and many smaller).
- Plates have formed from either continental or oceanic crust.
- They move over the semi-molten layer of the mantle (the **asthenosphere**).
- **Convection** and **slab-pull ridge-push** processes move the plates relative to each other at different speeds and in different directions.
- This creates:
  - divergence – plates pulling apart (**sea floor spreading** and rifting) at constructive boundaries
  - convergence – plates moving together (**subduction** and collision) at destructive and collision boundaries
  - transform movement – plates sliding past each other (fault lines) at conservative margins.

REVISION TIP

The concepts and processes in this section are all linked. Try revising this page once and then go back and try again. The second time will be much easier!

Continental and oceanic crust

Continental crust	Oceanic crust
- Thicker (35 to 40 km) - Less dense rock (e.g. granite) - Older – formed earlier in Earth's history (SiAl rocks, meaning silica and aluminium)	- Thinner (6 to 10 km) - Denser rock (e.g. basalt) - Younger – formed more recently at **plate boundaries** (SiMa rocks, meaning silica and magnesium)

Tectonic plate processes

Convection	Slab-pull ridge-push
Earth's core sends plumes of superheated mantle upwards. These rise to the asthenosphere, where they move plates through friction. The slightly cooled mantle sinks back down, creating a convection cell.	The density and gravity pull of the subducting plate pulls the rest of the plate downwards. On the other side of this plate, a mid-ocean ridge, higher than the rest of the plate, pushes the plate outwards.

Sea floor spreading	Subduction
Plates diverge as mantle rises and cools to form new oceanic crust, creating mid-ocean ridges between two oceanic plates. It can also create rift valleys between continental plates.	Plates converge and denser oceanic crust descends (subducts) under continental crust into the asthenosphere. The oceanic plate melts under heat and pressure, and a zone of earthquakes is created along the line of subduction.

⚙ KNOWLEDGE

1   Tectonic processes and hazards

1.1   Plate tectonics

Palaeomagnetism

This occurs when molten, magnetic minerals align themselves to Earth's magnetic field. When the rock solidifies, they are fixed as a fossil compass within the rocks. As Earth's magnetic field reverses its polarity every 300,000 years, scientists use the rocks to see how the crust has developed.

- Evidence from sea floor spreading shows the sea floor getting younger the nearer it is to a mid-ocean ridge.
- Pole-wandering looks at the fossil compass within rocks in continents and suggests they have drifted.

REVISION TIP

Palaeomagnetism is complex. Try to focus on its *use as evidence* rather than how it works.

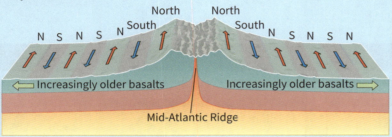

◀ **Figure 3** *Palaeomagnetism provides evidence of sea floor spreading*

Constructive plate margins

- Oceanic or continental plates move apart (diverge).
- Magma is forced up toward lower pressure areas.
- Magma solidifies to create new oceanic plate.
- Ocean ridges (and islands) form between oceanic plates.
- Rift valleys form between continental plates.
- Volcanoes (shield) and earthquakes (shallow focus) occur.

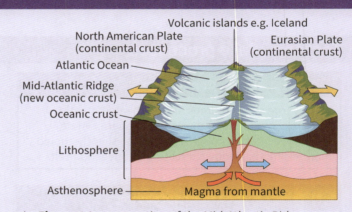

▲ **Figure 4** *A cross-section of the Mid-Atlantic Ridge, which forms the boundary between the Eurasian and North American Plates*

Conservative plate margins

- Created when the strain of movements makes plates move sideways relative to one another (transform).
- A fault line is created on the line of movement.
- No crust is created or destroyed, so no volcanoes occur. But earthquakes are common (shallow focus).

▶ **Figure 5** *A conservative plate margin*

Destructive plate margins

- Oceanic plates move toward (converge) and subduct under continental plates or less dense oceanic plates.
- Oceanic plates melt under heat and pressure; molten rock rises in plumes to form volcanoes.
- Ocean trenches form where the oceanic plate descends, pulling the other plate down.
- Volcanoes occur in chains on one side of the plate boundary, above the subduction zone, and earthquakes (shallow to deep focus) occur at the Benioff zone.

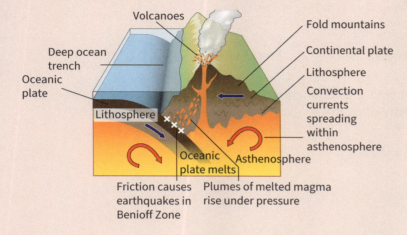

◀ *Figure 6 When oceanic plate meets continental plate*

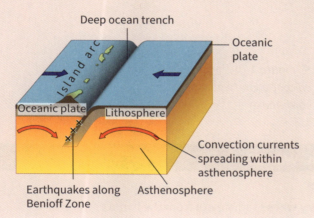

◀ *Figure 7 When oceanic plate meets oceanic plate*

Collision plate margins

- Continental plates move towards each other (converge).
- No subduction occurs (because both plates are low density, similar crust) but the collision causes crust to crumple and creates mountain ranges.
- No volcanoes occur, but earthquakes are common (shallow focus).

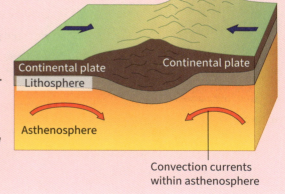

▶ *Figure 8 When continental plate meets continental plate*

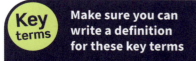

Key terms Make sure you can write a definition for these key terms

asthenosphere continental drift convection lithosphere
palaeomagnetism plate boundaries sea floor spreading
slab-pull ridge-push subduction

1   Tectonic processes and hazards

1.2 Earthquakes

Physical processes causing earthquakes

Earthquakes are when plate movements release energy as shaking in Earth's crust.

- Plates continually moving against each encounter friction.
- Pressure from this friction builds until it is suddenly released as different **seismic waves** of energy, which we feel as shaking.
- The earthquake's seismic waves radiate out from the point where the plate has slipped (**hypocentre**) and the point directly above this on the ground surface (**epicentre**).

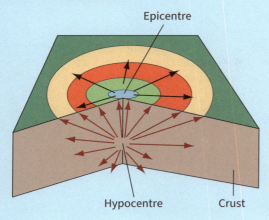

◀ **Figure 1** The hypocentre, epicentre, and waves of an earthquake

Where earthquakes occur

- At **fault lines**, where plates sliding past each other create a line of fractures which lock together until an earthquake releases the pressure.
- At a destructive boundary where oceanic plates are being subducted, creating a line of earthquakes at the **Benioff zone**.
- Intra-plate earthquakes can also occur far away from a plate boundary. This is due to smaller fault lines extending into the centre of plates.

Seismic waves

P waves:

- **Primary waves** radiate out from the hypocentre.
- They are the fastest waves and reach a location first, although are less damaging.
- They are compressional waves with a back-and-forth motion.

S waves:

- **Secondary waves** also radiate out from the hypocentre.
- They are slower than P waves and more damaging.
- They are transverse waves with a side-to-side motion from the hypocentre.

L waves:

- **Love waves** radiate out from the epicentre.
- They are the slowest waves and the most damaging, although they do not travel far.
- They move back and forth on the surface from the epicentre.

> **REVISION TIP**
>
> You need to know the type of movement and the degree of impact of the types of seismic wave. Try to create a mnemonic to help you remember this information.

Earthquake hazards

Primary hazards

Primary hazards are the direct impacts from a hazard.

- Crustal fracturing: the ground splits, which can cause building and infrastructure damage.
- Ground shaking: the ground shakes due to seismic waves. This can destroy buildings and infrastructure.

REVISION TIP

These hazards are generic. Make sure you remember to revise the examples and details from the case study you have studied.

Secondary hazards

Secondary hazards are a consequence of the primary hazards.

- **Liquefaction:** saturated soils can behave like water during shaking, leading to building foundations and infrastructure sinking.
- Landslides: shaking activates loose material on slopes, triggering landslides (these can include mudslides and rock falls).

Tsunami hazards

- **Tsunamis** may be seen as secondary hazards.
- They are extremely damaging due to the flooding of successive powerful waves and human activity on coasts.

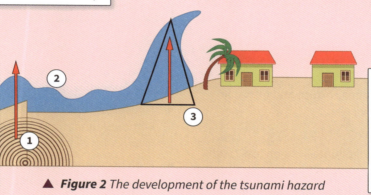

2. The displaced water becomes large waves, which ripple outward at immense speed. In deep oceans, the waves are low but deep.

1. Tsunamis are waves generated by submarine earthquakes at subduction zones. The sea floor rising or falling displaces a large amount of water directly above it (a water column).

3. As they approach the shore, the waves become higher as the sea floor becomes shallow. They can be channelled into bays due to local topography.

▲ **Figure 2** The development of the tsunami hazard

REVISION TIP

The AO1 part of an answer requires you to demonstrate accurate and relevant geographical knowledge and understanding. This means the retrieval of detailed examples.

 Key terms Make sure you can write a definition for these key terms

Benioff zone   epicentre   fault line   hypocentre   liquefaction   Love wave   primary wave   secondary wave   seismic waves   tsunami

1 Tectonic processes and hazards

1.2 Earthquakes

Tohoku, Japan 2011: a tsunami affecting a developed country

A magnitude 9.0 earthquake struck 100 km east of Sendai in Japan. It resulted in a tsunami up to 10 m high, travelling up to 10 km inland.

Social impacts	Economic impacts	Environmental impacts
• 20,000 dead or missing • 470,000 people made homeless initially	• Huge areas of Sendai destroyed • Total damage estimated at US$360 billion	• Radioactivity released from the Fukushima Daiichi Nuclear Power Plant • 500 km² of land flooded with seawater, affecting agriculture

Haiti 2010: an earthquake affecting a developing country

In January 2010, a 7.0 magnitude earthquake struck Haiti near the capital Port-au-Prince. It was triggered by movements at the boundary between the Caribbean and North American plates.

Vulnerability was very high:

- Heavy debts.
- No disaster planning from a corrupt and unstable government, close to civil war.
- Up to 90% unemployment.
- High poverty rate: 70% of people lived on less than $2 a day.
- Basic infrastructure and services like water supply didn't exist and there was very little access to clean water (33%).

REVISION TIP

If there are too many facts to retrieve, just aim for one from each of the social, economic, and environmental impacts.

- No building controls or regulations led to dangerous and poor informal housing.
- Housing in informal settlements were constructed without foundations, on steep slopes and soft ground. Large areas of this housing collapsed easily.
- High population growth and urbanisation led to huge, densely populated neighbourhoods without infrastructure.
- Low education rate: only 35% of population could read.

Social impacts	Economic impacts	Environmental impacts
• 220,000 deaths • 3.5 million people affected • 300,000 homes destroyed or damaged • Cholera outbreak	• US$14 billion in costs • 65% of Haiti's economic activity in the Port-au-Prince area that was severely affected • 8.5% of all jobs lost	• Landslides occurred • Coral reefs raised above sea level

China 2008: an earthquake affecting an emerging country

In May 2008, a 7.9 magnitude earthquake struck the mountainous area of Sichuan in China, triggering landslides.

Social impacts	Economic impacts	Environmental impacts
• 88,000 deaths (one third by landslides) • 45.5 million people affected • 4 million homes destroyed	• Estimated US$86 billion in damages • 14,000 industrial companies damaged (e.g. Dongfang Electric)	• 56,000 landslides triggered across the region • Rivers blocked and ecosystems disrupted

1.3 Volcanoes

Physical processes causing volcanic eruptions

Volcanoes are breaks in Earth's crust where magma funnels up due to heat and pressure, and escapes as lava, ash, and gas. Volcanoes occur:

- at destructive boundaries, formed from magma derived from subducted oceanic plate (intermingled with sedimentary rock and water) rising through a continental plate
- at constructive boundaries, formed of magma derived from the asthenosphere
- at **hotspots**, formed of magma derived from the asthenosphere which then rises through a plate.

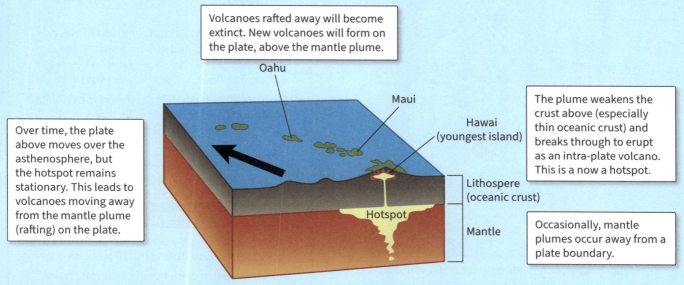

Volcanoes rafted away will become extinct. New volcanoes will form on the plate, above the mantle plume.

Over time, the plate above moves over the asthenosphere, but the hotspot remains stationary. This leads to volcanoes moving away from the mantle plume (rafting) on the plate.

The plume weakens the crust above (especially thin oceanic crust) and breaks through to erupt as an intra-plate volcano. This is a now a hotspot.

Occasionally, mantle plumes occur away from a plate boundary.

Oahu · Maui · Hawai (youngest island) · Lithospere (oceanic crust) · Hotspot · Mantle

▲ **Figure 1** *Hotspots form over mantle plumes and plate movement creates island chains, such as Hawaii*

Types of magma

The type of magma forming a volcano can lead to a significant difference in the magnitude and type of eruption, as well as the type of volcano formed.

Explosive eruptions
- High magnitude
- Destructive boundary
- Cooler **acidic lava** (still 1000°C)
- More viscous lava, moves slowly
- Bubbles of gas
- Creates smaller composite volcanoes

Gentle eruptions
- Low magnitude
- Constructive boundaries and hotspots
- Hotter **basic lava** (1500°C)
- Less viscous lava, moves quickly
- Dissolved gases
- Creates large shield volcanoes

REVISION TIP

The type of eruption and type of magma are linked. Try to prioritise this for your retrieval.

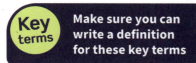 **Key terms** Make sure you can write a definition for these key terms

acidic lava   basic lava   hotspot   jökulhlaup   lahar
lava flow   pyroclastic flow   tephra   volcano

1 Tectonic processes and hazards

1.3 Volcanoes

Volcanic hazards

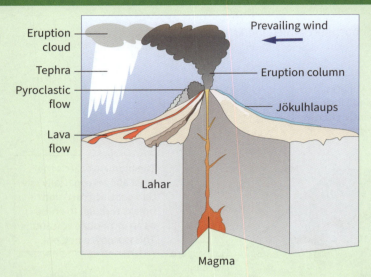

Eruption cloud
Tephra
Pyroclastic flow
Lava flow
Lahar
Magma
Prevailing wind
Eruption column
Jökulhlaups

▲ **Figure 2** *Primary and secondary volcanic hazards*

> **REVISION TIP**
>
> Jökulhlaups can also be described simply as glacial floods.

Primary hazards

- **Lava flows**: streams of lava that may move slowly but continually. They are very hot and may start fires.
- **Pyroclastic flows**: a combination of lava, ash, and gas which explodes from a volcano and travels at speed (100 km per hour). They are usually only found at acidic lava eruptions.
- **Tephra** and ash falls: large pieces of rock (tephra), lava bombs, and ash are thrown high into the atmosphere and fall back to Earth to deeply cover the land.
- Gas eruptions: gas is released from magma and may be hazardous if poisonous.

Secondary hazards

- **Lahars**: ash falls may combine with rainfall or snowmelt to form deadly mudflows called lahars.
- **Jökulhlaups**: in colder countries, volcanoes often have glacial ice above them. Eruptions melt the ice in huge quantities leading to glacial outburst floods called jökulhlaups.

Iceland 2010: a volcanic eruption affecting a developed country

In April 2010, Eyjafjallajökull erupted, leading to a jökulhlaup and ash cloud across southern Iceland.

Social impacts	Economic impacts	Environmental impacts
• No deaths • 800 people evacuated • Some had breathing issues	• Sheep, cattle, and horse farming affected by ash at 120 farms • A drop in tourism for months then a 16% rise above pre-eruption levels	• A jökulhlaup temporarily flooded a large area • A huge plume of ash sent 9 km into the skies around the North Atlantic

Montserrat 1995 to present – volcanic eruptions affecting a developing country

In 1995, the Chances Peak volcano erupted, beginning a series of explosive eruptions on the small island of Montserrat in the Caribbean. Pyroclastic flows were compounded by heavy ash falls and lahars.

Social impacts	Economic impacts	Environmental impacts
• 19 deaths by 1997 • Two thirds of homes, including most of the capital city, Plymouth, destroyed • Younger people emigrated away	• Tourism collapsed, leading to unemployment • Farmland destroyed or made inaccessible	• Gas and ash caused acid rain • Huge areas covered by ash, pyroclastic flows, lahars

RETRIEVAL

Learn the answers to the questions below, then cover the answers with a piece of paper and write as many as you can. Check and repeat.

Questions | Answers

#	Question	Answer
1	Name the layer of Earth upon which the lithosphere sits.	The asthenosphere
2	Name the two processes which move tectonic plates.	Convection and slab-pull ridge-push
3	What is the name of the process where the sea floor diverges and new plate is formed?	Sea floor spreading
4	Give two characteristics of oceanic crust.	Two from: less dense / thinner / younger / made of SiMa rich rocks
5	Give the name of the plate boundary where subduction occurs.	Destructive
6	Name the location on Earth's surface closest to the hypocentre of an earthquake.	Epicentre
7	Give two characteristics of S waves.	Two from: slower than P waves / more damaging than P waves / side-to-side movement
8	What is liquefaction?	Where saturated soils can behave like water during an earthquake
9	What name is given to the zone of earthquakes produced during subduction?	Benioff zone
10	What creates a tsunami?	A submarine earthquake displaces a column of water as the sea floor moves
11	At which plate boundaries do volcanoes form?	Destructive and constructive (also hotspots)
12	Give three characteristics of explosive magma.	Three from: hotter / less viscous / more dissolved gas / basic (basaltic) / creates shield volcanoes
13	What is a pyroclastic flow?	A combination of lava, ash, and gas that explodes at speed during a volcanic eruption
14	What happens if a volcano erupts under a glacier?	A secondary hazard known as a jökulhlaup (glacial flood) occurs
15	What elements create a lahar?	A mixture of ash fall / tephra and rainfall or snowmelt
16	State one social impact of the earthquake in Haiti in 2010.	One from: 220,000 deaths / 3.5 million people affected / 300,000 homes destroyed and damaged / Cholera outbreak
17	State one social impact of the Montserrat eruption in 1995.	One from: 19 deaths / two thirds of homes destroyed / most of Plymouth destroyed / younger people emigrated
18	State one economic impact of the Eyjafjallajökull volcanic eruption in 2010.	One from: sheep, cattle, and horse farming affected by ash at 120 farms / a drop in tourism for months then a 16% rise above pre-eruption levels

Put paper here

1 Tectonic processes and hazards

1.4 Vulnerability and disasters

The hazard risk equation

- A natural hazard is a natural event that could potentially threaten life and property.
- A **disaster** occurs when it actually causes impacts (the realisation of this hazard).
- For a disaster to occur, there needs to be a vulnerable population who can be affected.

Hazards and vulnerability combine for us to understand risk through the **hazard risk equation**, where capacity to cope means a community's **resilience** in coping with a hazard:

$$\text{Risk (R)} = \frac{\text{Hazard (H)} \times \text{Vulnerability (V)}}{\text{Capacity to cope (C)}}$$

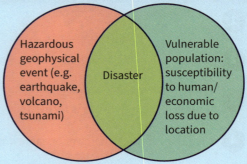

▲ *Figure 1* Degg's model

Factors that increase vulnerability

- **Inequality:** unequal access to housing, education, health, and jobs can increase vulnerability and reduce resilience after a disaster.
- Governance: a poorly governed country has much greater vulnerability through a lack of planning, regulations, and **preparedness** for disasters.

- Population factors: older people, women, and children are disproportionately affected by disasters.
- Physical factors: highly populated areas, rapid urbanisation, and low accessibility can increase vulnerability.

Pressure and Release (PAR) model

This suggests a progression of vulnerability factors, where root causes lead to dynamic pressures which lead to unsafe conditions.

The model can be used in hazard management by looking to 'release' the pressure and reduce disaster risk by dealing with the vulnerability factors.

The combination of these pressures with hazards explains why some countries are so vulnerable to natural hazards and prone to disasters.

```
Root causes  →  Dynamic pressures  →  Unsafe conditions  →  ✹ Disaster  ←  Hazards
```

- Limited access to power and resources.
- Poor political and economic systems.

- Lack of skills, training, and investment.
- Lack of free press.
- Rapid population growth, urbanisation, and deforestation.
- High debt.

- Fragile environment, economy, and society.
- Poor quality, low-income housing.

REVISION TIP

What the PAR model really indicates is that the relationships progress, must include both hazards and vulnerability, and are complex.

◀ *Figure 2* A disaster PAR model

Impacts of hazard events and vulnerability

The social and economic impacts of hazard events across the world vary depending upon the hazard itself and the vulnerability in developed, emerging, and developing countries.

Haiti 2010

An earthquake affecting a developing country.

Vulnerability was very high:

- Widespread poverty meant basic infrastructure and services like water supply didn't exist.
- Large areas of housing in informal settlements easily collapsed.
- No disaster planning from a corrupt and unstable government.

China 2008

An earthquake affecting an emerging country.

Vulnerability was lower:

- Disaster plans were immediately enacted with provision of aid and temporary housing.
- Healthcare was good and disease outbreaks prevented.
- However, bribes made to ignore regulations meant many buildings still collapsed.

Indian Ocean 2004

A tsunami affecting developing and emerging countries.

Vulnerability was relatively high:

- The areas affected were developing or emerging countries with a lack of warning, defences against tsunamis, or well-developed responses to hazards.
- High population densities on the coast often accompanied by high deforestation of protective mangroves.
- Fragile economies heavily reliant upon fishing and/or tourism.

Tohoku, Japan 2011

A tsunami affecting a developed country.

Vulnerability was relatively low:

- Clear disaster planning, including warning sirens and phone alerts, marked evacuation routes, and 10 m-high tsunami walls.
- Strict building regulations meant homes were built to withstand tsunamis and earthquakes.
- However, even some tsunami-proof buildings collapsed and older people couldn't get to safety in time. Of those who died, 66% were over 60.

Montserrat 1995 to present

Volcanic eruptions affecting a developing country.

Vulnerability was reasonably high:

- The country was low income (GDP of US$8253 in 1995), meaning a lack of preparation for eruption or emergency relief.
- The country is isolated and many areas inaccessible, leading to poor communications.
- However, international aid, troops, and monitoring made the eruption more predictable and supported the victims.

Iceland 2010

A volcanic eruption affecting a developed country.

Vulnerability was low:

- High-tech monitoring and forecasting of the eruption, meaning evacuation and warnings, previously practised in a full-scale live evacuation exercise in 2006.
- Well-developed emergency services and disaster plans in place from the Icelandic Department of Civil Protection and Emergency Management (ICP).

REVISION TIP

Look back at the examples in 1.2 and 1.3 and remind yourself of the impacts of each hazard event while you read about the level of vulnerability of each location.

REVISION TIP

Try to use the same examples for different parts of the course. You will have less to remember.

REVISION TIP

If you have learned different examples, add a few details to this page on a sticky note.

 Key terms Make sure you can write a definition for these key terms

disaster   hazard risk equation   inequality   preparedness   Pressure and Release (PAR) model   resilience

1 Tectonic processes and hazards

1.5 Measuring hazards

Measuring earthquakes

Earthquake events can be recorded using **seismographs** to measure the shaking of the ground during the period of an earthquake. This produces a seismogram trace, which can be used to analyse the waves involved at a location.

The magnitude (size or extent) and intensity (effects) are measured using different scales:

- **Richter scale**: measures magnitude through the height of seismic waves (amplitude). It is a logarithmic scale, increasing by a factor of ten at each 0–9 step.

- **Mercalli scale**: measures the effects through damage caused and what people experience in an earthquake on a I–XII scale.

- **Moment magnitude scale (MMS)**: measures the magnitude of energy released from an earthquake by combining the size of the waves, the amount of fault slip (movement), and the nature of the fault itself to give the most accurate measurement of magnitude. This scale is used most by scientists.

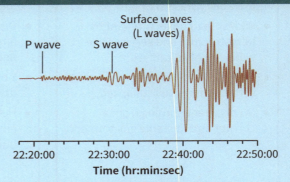

▲ **Figure 1** A seismogram measures the degree of shaking in earthquakes

REVISION TIP

You need to know how each scale works overall rather than every individual stage.

Measuring volcanic eruptions

The magnitude of a volcanic eruption can be recorded using various equipment:

- Tiltmeters: record how the surface of Earth is changed by the rising magma.

- Gas sampling: monitors gas release (e.g. sulphur).

- Seismographs: monitor the associated earthquakes.

- Tephra thickness: taken after the event to measure the thickness and nature of tephra deposits.

- Thermal imaging: using satellite or hand-held cameras to monitor heat from an eruption.

The magnitude of an eruption can also be measured using the **Volcanic Explosivity Index (VEI)**. This categorises the explosivity of eruptions on a logarithmic 0–8 scale, using qualitative terms to describe eruptions from 'gentle' to 'cataclysmic'. It considers the:

- height of the ash plume

- amount of tephra ejected

- duration of the eruption.

A higher VEI is associated with acidic lava and is much rarer, although it may have a more significant impact.

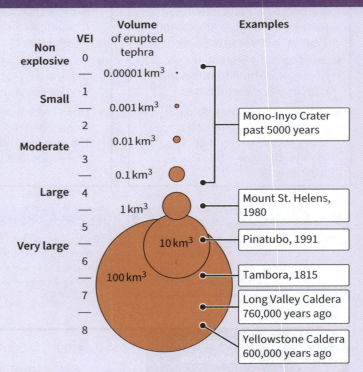

▲ **Figure 2** The VEI uses a logarithmic scale to record volcanic explosivity

Hazard profiles

A **hazard profile** is used to compare the physical characteristics of earthquake, volcanic eruption, or tsunami events. They can be used to develop management plans. Slower, smaller, and more predictable events are much easier to plan for than faster, larger, and more unpredictable ones.

The impacts from events can also be added to a hazard profile for comparison. A more severe event will usually score more highly for impacts.

▶ **Figure 3** *A hazard profile combined with impacts*

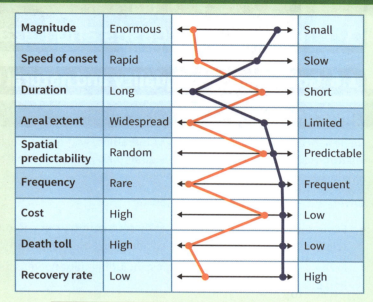

Magnitude	Enormous		Small
Speed of onset	Rapid		Slow
Duration	Long		Short
Areal extent	Widespread		Limited
Spatial predictability	Random		Predictable
Frequency	Rare		Frequent
Cost	High		Low
Death toll	High		Low
Recovery rate	Low		High

Key
● 2004 Indian Ocean tsunami ● Kilauea eruption

Magnitude: the size of an event. A larger size or scale of event is more dangerous.

Speed of onset: how quickly it occurs. A faster event gives less time to prepare and evacuate and is more dangerous.

Areal extent: the scale of an event. A larger area affected is usually more damaging.

Characteristics of hazard profiles

Duration: the length of time it lasts. Longer events are not always more dangerous.

Frequency: how often it occurs. Less frequent events are usually higher magnitude and more dangerous.

Analysis of magnitude and impacts

There is a positive correlation between magnitude and impacts such as death toll or damage.

A statistical test, such as the **Spearman's rank correlation coefficient**, is a technique used to analyse relationships. It tests the strength of correlation and gives a final number which is checked for the confidence of correlation. The correlation can also be shown a scatter graph.

REVISION TIP

Make sure you have practised completing a Spearman's rank test. In an exam, you will not need to remember the formula.

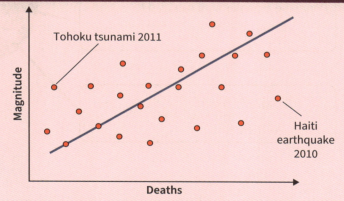

▲ **Figure 4** *A scatter graph of earthquake magnitude and deaths*

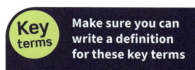

Key terms Make sure you can write a definition for these key terms

hazard profile   Mercalli scale   moment magnitude scale (MMS)   Richter scale   seismograph   Spearman's rank correlation coefficient   Volcanic Explosivity Index (VEI)

 **KNOWLEDGE**

1   Tectonic processes and hazards

1.6 Governance, inequality and natural hazards

Governance

Governance is a key factor affecting vulnerability and resilience. It involves the institutions, organisations, and individuals who manage people's lives through making decisions.

Governance can focus upon the following:

- Economics: making decisions that direct a country's economic activities and its relationship with other economies affecting equality, poverty, and quality of life.
- Politics: the formulation of policies, including national disaster reduction and planning.
- Administration: ensuring the policies above are implemented (e.g. enforcement of **building codes** or land-use planning).

Weak governance

Inequality of access to education, housing, and healthcare, and income can stem from poor governance, and all influence vulnerability and resilience to events.

Unstable governments mean decisions on disaster planning are not made.

High **corruption** means a country lacks investment in resources and infrastructure.

Poor land management can lead to deforestation, increasing the risk from landslides.

Building control regulations may be lacking or not be enforced, leading to informal settlements.

Risks increase with weak governance

Poor accessibility and isolated communities make it harder to make contact after a hazard strikes.

High population densities put more people in danger and increase the risk of disease.

Inequality is often high, meaning:
- Housing quality is poor and cannot withstand hazards.
- Education access is poor, with high levels of illiteracy and low levels of skills or training.
- Poor access to healthcare means a population may be less resilient and more likely to die of injuries or of disease that breaks out.

Contrasting hazard events in context

Haiti	**China**	**Japan**
Weak governance massively increased vulnerability leading up to the 2010 earthquake and reduced resilience afterwards.	Better governance reduced vulnerability and increased resilience leading up to the 2008 Sichuan earthquake.	Strong governance greatly reduced vulnerability and increased resilience leading up to the 2011 earthquake and tsunami.

Haiti
- Corrupt governments meant no disaster planning and huge waste alongside massive inequality.
- Governments allowed housing without foundations (informal settlements) to rapidly expand in densely populated areas (306 people per km² in the capital) with poor access.
- Weak governance meant infrastructure, healthcare, and education was extremely poor, with a literacy rate of 35%. Only 24% of people had access to sanitation.

China
- The strong and stable government meant disaster plans were immediately enacted, with 130,000 troops sent to provide aid.
- Temporary homes were set up for the homeless and US$10 billion provided for rebuilding.
- Healthcare was good and disease outbreaks prevented.
- However, bribes made to ignore building codes meant many buildings still collapsed.

Japan
- Building codes were well-engineered and strictly enforced, meaning relatively few buildings collapsed and defences protected against tsunami waves.
- Evacuation plans were enacted with text and siren warnings directing people to well-prepared emergency shelters. People had conducted drills in preparation.
- Trained troops were mobilised to provide aid, housing units were set up, and the economy was protected with loans.

REVISION TIP
These details show why Haiti was vulnerable. Try to recall one fact and expand it into a reason for vulnerability.

REVISION TIP
Try to remember only one factor to start with. Retrieval is a process which builds over time.

Contrasting statistics

The impact of governance and development level can mean that similar magnitude events produce significantly different outcomes.

	Haiti 2010	**China 2008**	**Japan 2011**
Type of country	Developing	Emerging	Developed
Magnitude of earthquake	7.0	7.9	9.0
Deaths	220,000	88,000	20,000
Homelessness	1.3 million	5 million	470,000
Cost	US$14 billion	US$86 billion	US$360 billion

REVISION TIP
If you can remember these numbers, they will give you core evidence of the impact of poor governance and inequality.

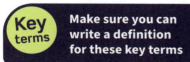 **Key terms** Make sure you can write a definition for these key terms

building codes   corruption

RETRIEVAL

Learn the answers to the questions below, then cover the answers with a piece of paper and write as many as you can. Check and repeat.

Questions

1 State the hazard risk equation.

2 Give one component part of the progression of vulnerability in the PAR model.

3 Give one environmental impact of the Tohoku tsunami, 2011.

4 What type of correlation is there between earthquake magnitude and impacts such as death toll or damage?

5 Give two social impacts of an earthquake affecting a developing country.

6 Give a scale used to measure earthquake magnitude.

7 Give two elements of the VEI scale.

8 Which two components of a hazard profile are missing from this list: speed of onset, areal extent, frequency, spatial predictability?

9 Define governance.

10 What type of settlement develops rapidly under weak governance?

Answers

Risk (R) =
Hazard (H) × Vulnerability (V) ÷ Capacity to cope (C)

One from: root causes / dynamic pressures / unsafe conditions

One from: radioactivity release / seawater flooding of farmland

A positive correlation

Two from: deaths / destruction of housing or homelessness / spread of disease

Richter scale or moment magnitude scale (MMS)

Two from: height of ash plume / amount of tephra ejected / duration of eruption

Magnitude and duration

The institutions, organisations, and individuals who manage people's lives through making decisions about the economy and laws, and ensuring rules are followed

Informal settlements

Put paper here

Previous questions

Now go back and use these questions to check your knowledge of previous topics.

Questions

1 What creates a tsunami?

2 Name the two processes which move tectonic plates.

3 What is liquefaction?

4 What name is given to the zone of earthquakes produced during subduction?

5 Name the layer of Earth upon which the lithosphere sits.

Answers

A submarine earthquake displaces a column of water as the sea floor moves

Convection and slab-pull ridge-push

Where saturated soils can behave like water during an earthquake

Benioff zone

The asthenosphere

Put paper here

1 Tectonic processes and hazards

1.7 Mega-disasters and multiple-hazard zones

Tectonic disaster trends

- The number of tectonic hazards has remained steady over time although other natural hazards have seemingly increased.
- The number of deaths fluctuate with spikes caused by mega-disasters.

- The number of people affected is fluctuating but rising slowly.
- The costs are fluctuating but rising steadily.

However, the data may not be fully **reliable**:

- Data, such as death tolls, may not be collected correctly in developing countries due to inaccessibility to remote areas.
- In the past, data was not collected at all in many places.
- Methods of data collection and definitions of death or damage differ between countries.
- Countries or NGOs may also want to inflate or reduce the figures for political reasons.

Mega-disasters

Mega-disasters are large-scale or large-impact events, which are often unpredictable and require significant responses. They have global or regional impacts:

- They may have direct impact upon a very large area, including more than one country.
- They may affect international supply chains and interdependent economies.
- The secondary environmental effects may cross borders after the event.

Tohoku earthquake and tsunami, 2011

This was a mega-disaster because:

- Radiation-contaminated water was released into the Pacific Ocean from the Fukushima Daiichi nuclear power plant, affecting regional fisheries.
- All nuclear power plants in Germany were closed due to the perceived risks following the Fukushima plant meltdown.
- Companies in the USA reduced production. For example, Boeing was supplied with engines by tsunami-affected IHI Corp and had to pause manufacturing.

Eruption of Eyjafjallajökull in Iceland, 2010

This was a mega-disaster because:

- Flights were cancelled across Europe for weeks due to the ash cloud, causing a US$2.6 billion loss in GDP and stranding passengers abroad for days.
- Companies (e.g. BMW) could not access components, meaning a halt in production.

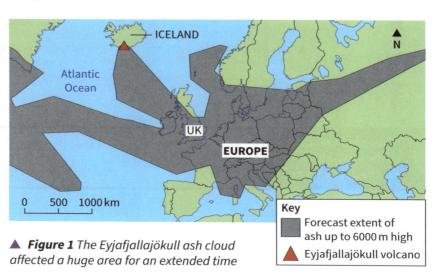

▲ **Figure 1** *The Eyjafjallajökull ash cloud affected a huge area for an extended time*

1 Tectonic processes and hazards

1.7 Mega-disasters and multiple-hazard zones

Multiple-hazard zones

- **Multiple-hazard zones** are locations which experience more than one hazard.
- Often, this is a combination of tectonic hazards with hydrometeorological (weather) hazards.
- Hazards can **interact** with each other to increase risks.
- They are sometimes known as **disaster hotspots** if combined with vulnerable populations, meaning a very high level of disaster risk.

> **REVISION TIP**
>
> You need knowledge of an example for both mega-disasters and multiple-hazard zones. The ones on these pages are those suggested by the specification. You may study others too.

The Philippines multiple-hazard zone

Floods are common due to monsoon rains, typhoons, and ENSO (El Niño), all of which are made worse by climate change.

The most tsunami-prone region in the Pacific.

Volcanoes are common across the destructive margin, with 20 active volcanoes (e.g. Mt Merapi).

100 to 150 earthquakes occur each year on the destructive margin of the Pacific and Philippine plates.

The mountainous landscape is prone to landslides, made worse by deforestation.

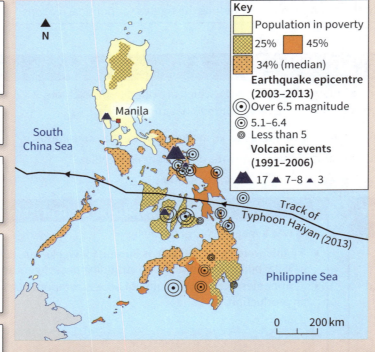

▲ *Figure 2 The vulnerability of the Philippines to natural hazards*

Interaction: Volcanic ash and typhoon rain mix together, creating lahars which are a bigger hazard than either single hazard (e.g. Mt Pinatubo and Typhoon Yunya 1991).

Interaction: Earthquakes and landslides combine, causing mudslides (e.g. in Leyte, 2006, where over 1000 people died).

Vulnerability: A developing country where 20% of the population lives in poverty, with rapid urbanisation and informal settlements called *barong barongs*.

Key terms Make sure you can write a definition for these key terms

disaster hotspot   disaster trend   interact   mega-disaster
multiple-hazard zone   reliable

1.8 Theoretical frameworks to understand hazards

Frameworks used

Frameworks used for **predication** (attempts to forecast), assessment of impacts, and deciding how to manage natural hazards include:

- **prediction** and **forecasting** methods
- the **hazard management cycle**
- **Park's model of disaster response**.

The hazard management cycle

- A simple model outlining four stages of management before and after a natural hazard strikes used by emergency planners (usually governments).
- It aims to reduce impacts and prepare for the next hazard, i.e. it is a cycle.
- The degree of recovery depends on the nature of the event and the level of development of the area, as well as governance.

Mitigation
Identification of potential hazards to minimise their impact through planning land use, building codes, and hazard defence structures. Mitigation can be costly and is less effective in developing countries.

Recovery
Rebuilding and getting back to normal over weeks to months. Includes restoring services, rebuilding homes, long-term medical care, and assisting the economy. This will show the level of short-term economic resilience in a country.

Response
The immediate help following a hazard – search and rescue, evacuation, emergency shelter, restoration of critical infrastructure (e.g. fresh water and medical care). An effective emergency response will save many lives in the short term.

Preparation
Pre-event activities focused upon communities. Includes evacuation planning, preparing aid, education, and attempts to predict hazards. Being aware and prepared for a hazard can greatly reduce the level of risk, and emergency planners are often used at this stage.

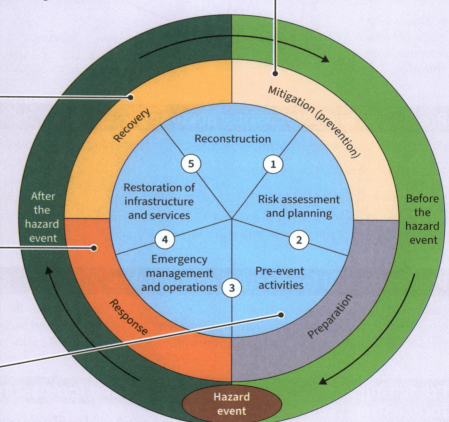

▲ *Figure 1* *The hazard management cycle*

 Key terms Make sure you can write a definition for these key terms

forecasting   hazard management cycle
Park's model of disaster response   predication   prediction

1 Tectonic processes and hazards

1.8 Theoretical frameworks to understand hazards

Park's model

Also known as the hazard-response curve, it can be used to compare the ability of different countries to recover from a hazard over time (resilience). Each country has its own curve.

▶ **Figure 2** *Park's model of hazard response (after Park 1991)*

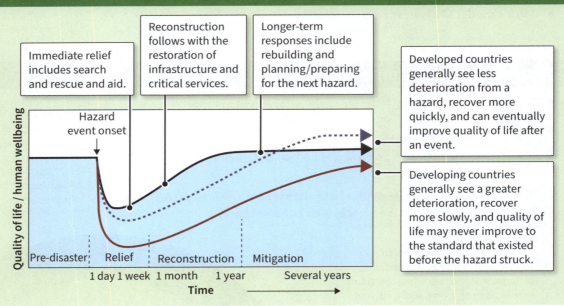

Immediate relief includes search and rescue and aid.

Reconstruction follows with the restoration of infrastructure and critical services.

Longer-term responses include rebuilding and planning/preparing for the next hazard.

Developed countries generally see less deterioration from a hazard, recover more quickly, and can eventually improve quality of life after an event.

Developing countries generally see a greater deterioration, recover more slowly, and quality of life may never improve to the standard that existed before the hazard struck.

Hazard event onset

Quality of life / human wellbeing

Pre-disaster | Relief | Reconstruction | Mitigation

1 day 1 week 1 month   1 year   Several years

Time

Prediction and forecasting accuracy

Prediction requires a high-certainty judgement, while forecasting includes much more uncertainty.

The accuracy of prediction depends upon:

- the type and location of a hazard
- the level of development and governance of the country involved (scientists in developed countries can predict events with much higher accuracy).

Predicting and forecasting earthquakes

Predicting earthquakes is effectively impossible. Forecasts can be made using the intervals between historic earthquakes in a location (recurrence intervals).

Intervals may be regular because an earthquake releases the strain on a plate boundary, which then steadily increases again.

Predicting and forecasting tsunamis

Only very short-term forecasts are made by analysing buoys for wave height following earthquakes (e.g. DART system in the Pacific) and calculating the degree of spatial risk using computer modelling, such as the Tsunami Early Warning Systems (TEWS). Scientists are key players designing and implementing these systems.

Predicting and forecasting volcanic eruptions

Monitoring volcanoes can detect changes indicating magma moving to the surface. Fairly accurate forecasts of eruptions can then be made. Monitoring includes measuring:

- water levels and temperatures, which may change with rising magma
- increased release of gases (e.g. radon or sulphur), which may indicate magma changes
- deformation of the volcanic cone – this occurs as magma rises, causing bulging; temperature changes can also be observed using satellite imaging
- small earthquake (microquake) activity, which increases as magma forces it way through the rock.

1.9 Hazard management

Strategies to manage tectonic hazards

Increasing technology →

Modify the loss	Modify vulnerability and resilience	Modify the event
Reduce the losses after the event. This is easy to put into place, but can be costly for developing countries. • Emergency and short-term aid • Longer-term aid • Insurance	Reduce the risk to people by reducing their vulnerability and increasing resilience before an event. This is easier and cheaper for developing countries, but requires good governance. • High-tech monitoring • Prediction • Education • Community preparedness and adaptation	Reduce the danger by changing the hazard itself before and during an event. This can be technologically difficult and expensive, so it has limited application to developing countries. • Land-use zoning • Hazard-resistant design • Engineering defences

These attempts can also be seen as ways to **mitigate** and **adapt** to the hazard risk. Attempts to modify the cause of the hazard are not feasible for tectonic hazards.

Strategies to modify the event

Land-use zoning: areas at highest risk are identified and mapped by planners using technology like GIS. These 'red zones' may have restrictions on the type of buildings (e.g. residential) and the height of buildings, and include evacuation routes.

Advantages	Disadvantages
☑ Easy to implement and may save many lives ☑ Low cost	✗ Requires high administrative governance to be successful ✗ Relies on modelling

Hazard-resistant design: buildings are designed to withstand tectonic hazards and older buildings can be retrofitted. Designs include cross-bracing and pendulum weights in high-rise blocks, shock-absorbing foundations to withstand seismic waves, and reinforced roofs for volcanic ashfalls.

Advantages	Disadvantages
☑ Prevents the event destroying buildings and infrastructure and reduces the need to rebuild ☑ Existing buildings can be retrofitted	✗ May be very expensive, so limited to developed countries ✗ Requires high technology ✗ Requires strong governance to be successful

Engineering defences: physical defences can be constructed, such as sea walls to withstand tsunamis. Walls like this in Japan are up to 15 m high. Attempts have also been made by engineers to divert lava flows using earth barriers. During 2001, for example, 13 barriers were built around Mt Etna's lava flows.

Advantages	Disadvantages
☑ Prevents the hazard from affecting infrastructure and buildings ☑ Reassures the public	✗ May be very expensive, so limited to developed countries ✗ High-magnitude events can still overtop or destroy defences

KNOWLEDGE

1 Tectonic processes and hazards

1.9 Hazard management

Strategies to modify vulnerability and resilience

High-tech monitoring and prediction: monitoring volcanoes or ocean height can provide prediction and forecasts to allow for evacuation. In Japan, all smartphones are activated to repeat '*Jishin desu! Jishin desu!*' meaning 'There is an earthquake!'.

Advantages	Disadvantages
☑ Provides time for emergency preparation and evacuation ☑ Monitoring is increasingly remote and forecasts based on more accurate models	☒ Forecasts are not always accurate and can create false alarms ☒ Requires high technology and trained scientists to be effective

Education: earthquake drills take place in many countries, including in Japan on 1st September, their 'Disaster Prevention Day'.

Advantages	Disadvantages
☑ Relatively cheap, especially for developing countries	☒ May be ineffective with large magnitude events ☒ Infrastructure and homes may still be damaged

Community preparedness and adaptation: having locally designed evacuation plans and local knowledge of the landscape and population means better preparation.

Advantages	Disadvantages
☑ Relatively cheap, especially for developing countries ☑ Provides specific planning for an area	☒ Reliant upon community involvement ☒ Harder to provide for in remote areas

Strategies to modify the loss

Aid: often provided by NGOs, it is especially important for less-resilient developing countries. Emergency aid provides the immediate essentials like food, water, medical care, and shelter. Short-term aid restores essential services like electricity supply. Longer-term aid rebuilds and supports the economy.

> **REVISION TIP**
>
> Players are noted in the specification, so you may be asked a question which requires your knowledge of their role in managing hazards.

Advantages	Disadvantages
☑ Provides immediate help to vulnerable populations ☑ Long-term aid helps reconstruction	☒ May be expensive ☒ Often reliant upon foreign donations ☒ Possible abuse by corrupt governments

Insurance: insurers provide economic cover for losses including infrastructure, homes, and losses incurred to business.

Advantages	Disadvantages
☑ Provides funds for reconstruction and future mitigation ☑ Prevents long-term damage to the economy	☒ Expensive, so limited to developed countries ☒ Does not prevent loss of life or damage to buildings and infrastructure

Actions of communities: local people are vital for search and rescue following a hazard event, especially in remote parts of developing countries.

Advantages	Disadvantages
☑ Relatively cheap, especially for developing countries	☒ Reliant upon community involvement

Key terms — Make sure you can write a definition for these key terms

adapt   engineering defences   hazard-resistant design
land-use zoning   mitigate

RETRIEVAL

Learn the answers to the questions below, then cover the answers with a piece of paper and write as many as you can. Check and repeat.

Questions

1. Define a mega-disaster.

2. Give one reason why the Eyjafjallajökull volcanic eruption in 2010 is classed as a mega-disaster.

3. Give one example of an interaction of hazards that occurs in the Philippines, a multiple-hazard zone.

4. Which component of the hazard management cycle comes after 'preparation'?

5. What type of countries tend to improve in quality of life following a hazard, according to Park's model?

6. Which tectonic hazard has high inaccuracy in forecasts?

7. Name one method used to increase the accuracy of forecasts of volcanic eruptions.

8. What is modified by hazard-resistant design?

9. Name an engineering defence used to mitigate the risk from hazards.

Answers

Put paper here

1. A large-scale or large-impact event which is often unpredictable and requires significant responses

2. One from: flights cancelled across Europe / companies not able to access components slowing production

3. One from: volcanic ash and typhoon rain mixing to cause a lahar / earthquakes and landslides combine, causing mudslides

4. Response

5. Developed

6. Earthquakes

7. One from: monitoring: water levels / release of gases / deformation of the volcano / temperature changes / small earthquake activity

8. The event

9. One from: tsunami sea walls / lava diversion barriers / building bracing / shock-absorbing foundations / pendulum weights

Previous questions

Now go back and use these questions to check your knowledge of previous topics.

Questions

1. What is the name of the process where the sea floor diverges and new plate is formed?

2. State the hazard risk equation.

3. Give two social impacts of an earthquake affecting a developing country

4. Give two elements of the VEI scale.

5. What happens if a volcano erupts under a glacier?

Answers

Put paper here

1. Sea floor spreading

2. Risk (R) = Hazard (H) × Vulnerability (V) ÷ Capacity to cope (C)

3. Two from: deaths / destruction of housing or homelessness / spread of disease

4. Two from: height of ash plume / amount of tephra ejected / duration of eruption

5. A secondary hazard known as a jökulhlaup (glacial flood) occurs

PRACTICE

Exam-style questions

1 Study **Figure 1a** and **Figure 1b**, which show the VEI and estimated casualties of eleven selected volcanic eruptions.

Eruption	Year	Estimated casualties	VEI
Nevado del Ruiz, Colombia	1985	25,000	3
Mont Pelée, Martinique	1902	30,000	4
Krakatau, Indonesia	1883	36,000	6
Tambora, Indonesia	1815	92,000	7
Lakagigar (Laki), Iceland	1783	9,000	4
Kelut, Indonesia	1586	10,000	5
Mount Pinatubo, Philippines	1991		6
Mount St. Helens, Washington	1980	57	5
Kilauea, Hawaii	1924	1	
Lassen Peak, California	1915	0	5
Mount Vesuvius, Italy	79	3,360	5

▲ **Figure 1a** *A table showing the VEI and estimated casualties of eleven selected volcanic eruptions*

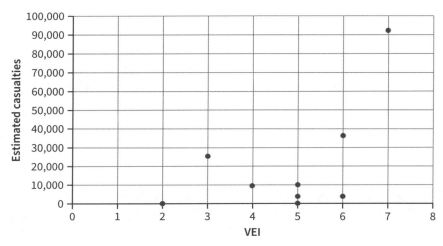

▲ **Figure 1b** *Graph showing VEI and estimated casualties of eleven selected volcanic eruptions*

(a) Using **Figure 1b**, complete **Figure 1a** by giving the correct estimated causalities for the Mount Pinatubo, Philippines 1991 eruption and the VEI for the Kilauea, Hawaii, eruption in 1924. **(2)**

(b) Complete **Figure 1b** by plotting the Mont Pelée, Martinique, eruption in 1902 using the data from **Figure 1a**. **(2)**

2   Study **Figure 2** which shows the average and median earthquake damage, 1985–2015, for selected earthquake events in developed countries.

Earthquakes in developed countries	Average damage ($mil, 2015 dollars)	Median damage ($mil, 2015 dollars)
All 5.5+ quakes	$12,146	$484
Magnitude > 6.5	$23,966	$628
Magnitude 5.5 to 6.5	$2,145	$178
Population > 250,000	$20,705	$1,980
Population < 250,000	$172	$28

▲ **Figure 2** *The average and median earthquake damage, 1985–2015, for selected earthquake events in developed countries*

(a)   Calculate the ratio of average damage ($mil, 2015 dollars) between Magnitude >6.5 events and Magnitude 5.5 to 6.5 events (to two decimal places).   **(2)**

(b)   What is the difference between median damage ($mil, 2015 dollars) in countries with populations <250,000 and countries with populations > 250,000?   **(1)**

(c)   Calculate the average damage per year ($mil, 2015 dollars) for all 5.5+ quakes (to two decimal places).   **(1)**

3   Study **Figure 3**.

This data was collected to investigate whether there was a significant relationship between the damage ($mil, 2015 dollars) and the magnitude of ten earthquakes.

Date	Location	Magnitude	Rank	Damage ($mil, 2015 dollars)	Rank	D	D_2
2011	Japan, Honshu	9.0	1	231,806	1	0	0
2003	Japan, Hokkaido	8.3	2	116	9	7	49
2002	Alaska	7.9	3	74	10	7	49
1994	Japan, Honshu	7.8	4	272	6	2	4
1993	Japan, Hokkaido/Russia	7.7	5	1,980	4	1	1
1993	Japan, Hokkaido	7.6	6.5	587	5	1.5	2.25
1992	USA, California (Yucca)	7.6	6.5	155	7	0.5	0.25
1992	USA, California (Humboldt)	7.1	8	127	8	0	0
2010	New Zealand	7.0	9	7,064	3	6	36
1989	USA, California (Loma Prieta)	6.9	10	10,708	2	8	64
						$\Sigma D_2 =$	

▲ **Figure 3** *Data collected to investigate whether there was a significant relationship between the damage ($mil, 2015 dollars) and the magnitude of ten earthquakes*

(a) Complete Figure 3 by calculating $\sum D_2$. **(1)**

(b) The formula for Spearman's rank correlation coefficient value r_s is given below. Here n is equal to 10. Calculate the value of r_s to three decimal places for the data given. You must show your working. **(2)**

$$r_s = 1 - \frac{6\sum D^2}{n(n^2 - 1)}$$

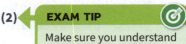

EXAM TIP

Make sure you understand how a Spearman's rank test works and always show your working!

(c) The critical value for a 95% confidence level is 0.564.

Null hypothesis: There is no significant relationship between the damage ($mil, 2015 dollars) and the magnitude of ten earthquakes.

Do you accept or reject the null hypothesis? **(1)**

4 Study **Figure 4** which shows the date, location and magniture of selected earthquakes.

Date	Location	Magnitude
2011	Japan, Honshu	9.0
2003	Japan, Hokkaido	8.3
2002	Alaska	7.9
1994	Japan, Honshu	7.8
1993	Japan, Hokkaido	7.6
1992	USA, California (Yucca)	7.6
1992	USA, California (Humboldt)	7.1
2010	New Zealand	7.0
1989	USA, California (Loma Prieta)	6.9

▲ **Figure 4** The date, location and magnitude of selected earthquakes

(a) Calculate the mean magnitude recorded. **(1)**

(b) Calculate the median magnitude recorded. **(1)**

(c) Calculate the interquartile range. Show your working. **(2)**

5   Study **Figure 5**.

The data was collected to investigate whether there was a significant association between the number of earthquakes and the depth of earthquakes in Turkey from 1950–2023.

	O	E	(O–E)	(O–E)²	(O–E)²/E
Depth	Number of earhquakes over magnitude 5	Number of earhquakes over magnitude 5			
0–10km	7	11	4	16	3.2
11–20km	23	11	12	144	13.1
21–30km	14	11	3	9	0.8
31–40km	5	11	–6		3.3
41–50km	6	11		25	2.3
					Σ =

▲ **Figure 5** *A χ^2 table to show the frequency and depth of earthquakes over magnitude 5 in Turkey from 1950–2023*

χ^2 = chi squared

O = Observed values

E = Expected values

(a)   Complete the table shown.   **(2)**

(b)   Calculate χ^2.   **(1)**

$$\chi^2 = \Sigma (O-E)2E$$

(c)   The critical vale for a 95% confidence level is 9.488.

Null Hypothesis: There is no significant association between the number of earthquakes over magnitude 5 and the depth.

Do you accept or reject the null hypothesis?   **(1)**

> **EXAM TIP**
>
> At A level, the four-mark question for this section is usually a statistic-based one. The context is not really important.

Exam-style questions

6   Study **Figure 6**.

Entity	Year	Earthquake deaths
China	1556	830,000
Haiti	2010	316,000
China	1976	242,769
Indonesia	2004	228,008
China	1920	200,000
Japan	1923	142,807
Turkmenistan	1948	110,000
China	2008	87,724
Azerbaijan	1667	80,000
Number of events: 9		
Total deaths: 2,237,308		

▲ **Figure 6** *The world's ten largest recorded earthquakes by number of deaths*

(a)   Calculate the mean number of deaths recorded.   **(1)**

(b)   Calculate the median number of deaths recorded.   **(1)**

(c)   Calculate the interquartile range. Show your working.   **(2)**

7   Study **Figure 7**.

Eruption and year	Deaths	Rank	VEI	Rank	D	D$_2$
Indonesia	250,000	1	7	2.5	−1.5	2.25
Indonesia	36,000	3	6	5	2	4
Martinique	30,000	5	4	7.5	2.5	6.25
Colombia	23,000	4	3	9	−5	25
Greece	20,000	6	7	2.5	3.5	12.25
Indonesia	15,000	7.5	7	2.5	5	25
Japan	15,000	7.5	2	10	−2.5	6.25
Italy	13,000	9	5	6	3	9
Iceland	10,000	10	4	7.5	2.5	6.25
Indonesia	71,000	2	7	2.5	−0.5	0.25
						$\sum D^2 =$

▲ **Figure 7** *Data collected to investigate whether there was a significant relationship between the number of deaths and the VEI magnitude of ten volcanic eruptions*

(a) The formula for Spearman's rank correlation coefficient value r_s is given below. Here, n is equal to 10. Calculate the value of r_s to two decimal places for the data given. You must show your working. **(2)**

$$r_s = 1 - \frac{6\sum D^2}{n(n^2 - 1)}$$

(b) The critical value for 0.05 confidence is 0.564. Do you accept or reject a null hypothesis? **(2)**

8 Assess the reasons why the impacts of tectonic hazards may vary between countries. **(12)**

EXAM TIP

The 12-mark questions in this section almost always require answers with the same component parts considering the reasons for differences in impacts.

9 Assess the significance of physical processes in explaining the impacts of tectonic hazards. **(12)**

10 Assess the importance of prediction and forecasting in managing the impacts of earthquake hazards. **(12)**

11 Assess the importance of governance in the management of disaster impact. **(12)**

EXAM TIP

An answer to a 12-mark 'assess' question doesn't necessarily need an introduction but does need a conclusion!

12 Assess the relative effectiveness of strategies used to modify the impacts of tectonic events. **(12)**

13 Assess the extent to which developing nations are often more vulnerable than others to tectonic hazards. **(12)**

14 Assess which stage of the hazard management cycle is the most important for managing the impacts of tectonic hazards. **(12)**

15 Assess the reasons why the number of people affected by tectonic hazards seem to be increasing. **(12)**

2.1 Causes of longer and shorter climate change

Pleistocene climate change

The most recent period of ice activity was the **Pleistocene** epoch which started 2.6 million years ago and ended almost 12,000 years ago. There were multiple (at least 10) **glacial periods** (leading to ice advances) and **interglacial periods** (causing ice retreats). Within each major glacial, there were also shorter fluctuations – known as **stadial** (colder) and **interstadial** (warmer) periods.

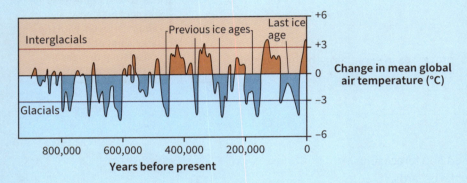

◀ **Figure 1** Global mean temperature fluctuations over the last million years.

◀ **Figure 2** Maximum ice extent during the Last Glacial Maximum (LGM) about 20,000 years ago. Ice sheets were 3–4 km in thickness.

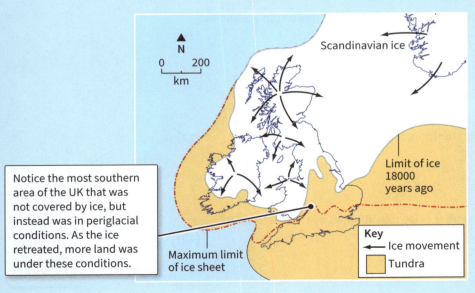

Notice the most southern area of the UK that was not covered by ice, but instead was in periglacial conditions. As the ice retreated, more land was under these conditions.

◀ **Figure 3** Extent of ice cover about 18,000 years ago, when the British Isles were joined to Europe

Loch Lomond stadial

During the end of the Pleistocene epoch (between 12,900 to 11,700 years ago) in the UK, the ice started to readvance, covering all the high land – including the West Highland Glacier Complex, and smaller ice fields in mountainous regions such as the Lake District – leaving moraine, ribbon and cirque lakes, as well as other glacial remnants. At the end of this period, there was rapid natural climate change, leading to full seasonal melting of all the ice.

Holocene climate change

Little Ice Age

Historically, global temperatures have been increasing. However, there have been smaller global coolings in recent history.

During the Middle Ages (950 to 1250), the climate was mostly fairly warm. This was followed by a Little Ice Age (also called the Maunder Minimum) lasting from around 1550 to 1850. Volcanic activity and less solar insolation caused a drop in the temperature that lasted for some time. Frost fairs were frequently held on the River Thames during the extra cold winters. It also led to food shortages and periods of famine in Europe and Asia.

Anthropocene climate change

- The **Anthropocene** is a proposed geological epoch from 1950 to present day (not ratified).
- Marks the significant human impact from population growth, pollution, increased exploitation of natural resources, and the start of the Atomic Age.

- Evidence of these can be found in sediments worldwide.
- This is a time of increased greenhouse gases in the atmosphere leading to anthropogenic (human-accelerated) climate change, e.g. elevated levels of carbon dioxide measured since the start of the Industrial Revolution.

Milankovitch cycles

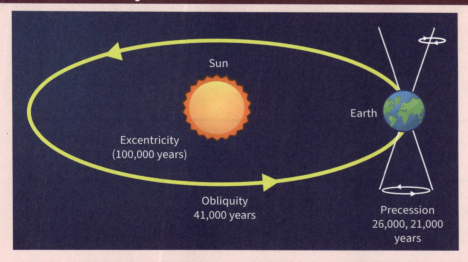

◄ **Figure 4 Milankovitch cycles** *are the primary cause of natural cyclical climate change*

	Cycle	Reasons	Impacts
Eccentricity	100,000 years	Periodic variations in the shape (eccentricity) of Earth's elliptical orbit around the Sun.	Changes the distance from the Sun over the cycle, and therefore the amount of solar radiation.
Obliquity (axial tilt)	41,000 years	Changes in the tilt of Earth's axis relative to its orbit around the Sun, ranging from about 22.1 to 24.5 degrees.	More extreme seasons, with warmer summers and colder winters in higher latitudes.
Precession	Axial precession ~26,000 years Orbital precession ~21,000 years	Wobbling of the Earth's axis: Axial precession (the slow rotation of the Earth's axis itself) Orbital precession (the slow change in the orientation of Earth's orbit).	Affects the timing of the seasons and can influence the intensity of the seasons in different hemispheres.

KNOWLEDGE

2 Glaciated landscapes and change

2.1 Causes of longer and shorter climate change

Variations in solar output

- Sunspots are temporary but cyclical phenomena on the Sun's photosphere that appear as dark spots compared to the surrounding areas.
- Associated with intense magnetic activity and often accompanied by solar flares and other solar events, creating a warmer climate on Earth.

Volcanic eruptions

- Large eruptions produce large volumes of carbon dioxide and other greenhouse gases.
- Can also produce large volumes of ash and sulphur dioxide, which can reflect solar radiation out of the atmosphere and lead to significant cooling of the climate on a temporary basis.

Variations in atmospheric gases

- Ice cores contain information about temperatures and atmospheric gas levels up to 800,000 years ago.
- Fossil fuel usage and changes in land-use have resulted in carbon dioxide levels that are now 50% higher than before the Industrial revolution. This increase is unprecedented in the last 800,000 years.
- The concentration of atmospheric methane and other greenhouse gases have also dramatically increased over the last two hundred years.

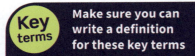 **Make sure you can write a definition for these key terms**

Anthropocene glacial period Holocene interglacial period interstadial Milankovitch cycles Pleistocene stadial

2.2 Past and present ice cover

The cryosphere and ice masses

The **cryosphere** refers to all frozen water on or under the Earth's surface (the hydrosphere). It interacts with all of Earth's natural systems through a series of feedback loops.

Ice masses

Ice sheets: enormous continental-scale masses of ice. Antarctica and Greenland are the only ice sheets present today.

Ice caps: smaller than ice sheets but still relatively large, dome-shaped and may cover entire islands or mountainous regions.

Glaciers: smaller masses of ice that form in mountainous regions. They flow downhill under the influence of gravity.

Icefields: intermediate in size between ice caps and glaciers, icefields are extensive areas of ice that cover mountainous terrain. They often feed several glaciers.

Polar ice: ice masses located near the Earth's poles. This includes the ice sheets of Antarctica and Greenland.

Temperate glaciers: located in regions where the temperature is warm enough for some melting to occur at the glacier's base. This results in the presence of liquid water within or at the base of the glacier.

Cirque (corrie/cwm) glaciers form in hollows in the highlands, slowly eroding them through plucking and abrasion.

Cirque glaciers eventually spill over the lip of the hollow to feed valley glaciers, which are larger and flow downhill eroding V-shaped glacial valleys into U-shaped valleys.

Present-day distribution of ice masses

- Above 66.7° latitude north and south, mostly in Antarctica and Arctic regions. Antarctica is colder due to being a large land mass.
- Upland regions (high altitude) Himalayas, Andes, Northern Rocky Mountains, and European Alps.

▶ **Figure 1** Present-day summer ice cover (permanent ice cover) in the Arctic region. Arctic sea-ice is melting fast due to climate change. The region is warming at least twice as fast as other regions.

Key
— Average extent of ice, 1981–2010

KNOWLEDGE

2 Glaciated landscapes and change

2.2 Past and present ice cover

Evidence for Pleistocene glaciation

Area	Evidence
Lake District	U-shaped valleys Moraines Drumlins
Scottish Highlands	U-shaped valleys Corries (cirques) Arêtes Hanging valleys Ben Nevis demonstrates erosional features
Snowdonia	Pyramidal peaks Cirques Moraines. Snowdon shows glacial erosion
Brecon Beacons in South Wales	U-shaped valleys Moraines
Cairngorms	Corries, arêtes, and glacial troughs, lochs and tarns
East Anglia	Drumlins Till Norfolk Broads show post-glacial isostatic rebound

REVISION TIP

Mnemonics are useful memory devices which can help you recall lists of causes and consequences, or a sequence of events or actions. Identify the keywords for the topic. Write them out. Make up a memorable sentence where each word has the same initial letters as the words in the sequence.

▲ **Figure 2** *A glaciated mountain corrie and lake in the Black Cuillin mountains, Isle of Skye, Scotland*

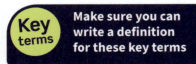

Key terms Make sure you can write a definition for these key terms

cirque / corrie / cwm   cryosphere   glacier
ice caps   ice sheets   temperate glacier

2.3 Periglacial processes produce distinctive landscapes

Periglaciation

- **Periglaciation** occurs in climates that are subject to repeated freezing and thawing. These areas are usually found on the margins of glaciated areas and cover approximately 25% of the global land surface.

- Due to climate change, many previously frozen areas are now subject to periglacial processes and periglacial areas are increasing as glacial areas decrease.

- Due to melting of snow and refreezing to ice, sparse vegetation and high winds, distinctive landforms are developed.

- Due to seasonal thawing, many areas have an overlying **active layer**, underlain by **permafrost** (ground that is continually frozen for at least two years).

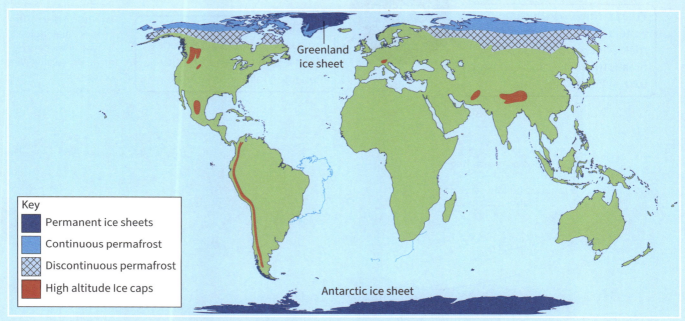

Key
- Permanent ice sheets
- Continuous permafrost
- Discontinuous permafrost
- High altitude Ice caps

▲ **Figure 1** *Current global distribution of cold environments*

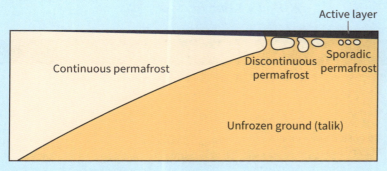

▲ **Figure 2** *Permafrost types*

 # KNOWLEDGE

2 Glaciated landscapes and change

2.3 Periglacial processes produce distinctive landscapes

Periglacial processes

Periglacial processes produce distinctive landscapes.

Nivation: erosion of the ground around a slope of snow due to freezing and thawing.

Freeze–thaw weathering: Water falls through cracks in rocks, freezes, expands and breaks the rock as it is exposed to the climate over time.

Periglacial processes

Frost heave: the uplift of soil due to the expansion of groundwater when it freezes.

Solifluction: the slow movement of wet soil/material down a slope, often forming elongated lobes of material.

REVISION TIP

Understand the difference between weathering and erosion. Weathering takes place 'in situ' (in place). In erosion, material is moved elsewhere by an agent – water, wind or ice.

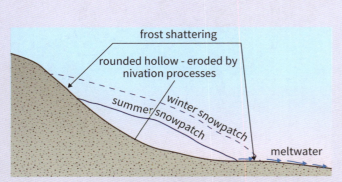

▲ **Figure 3** *Nivation processes that result in the formation of a nivation hollow*

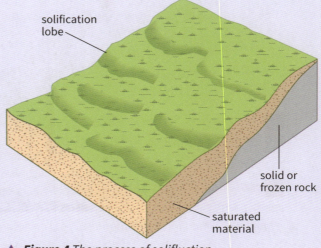

▲ **Figure 4** *The process of solifluction*

Periglacial landforms

The periglacial landforms created by periglacial processes create distinctive periglacial landscapes. There is often little vegetation.

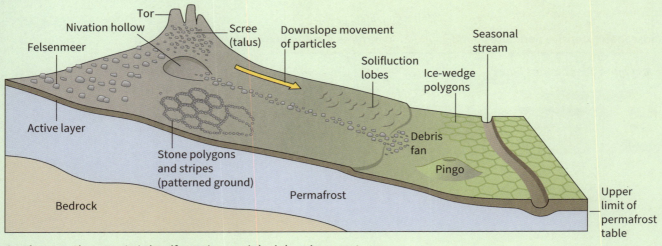

▲ **Figure 5** *Characteristic landforms in a periglacial environment*

Formation of periglacial landforms

Ice wedges

Water can melt into cracks and joints in the rock. As it freezes, it expands and forces sediment towards the surface. This process often produces patterned ground through frost heave.

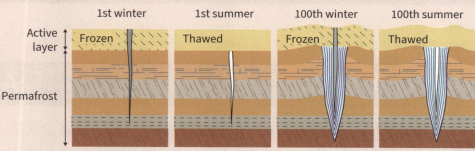

▲ **Figure 6** *Formation of ice wedges*

Patterned ground

- Repeated freezing and thawing cycles create patterns of sorted stones on the ground surface.
- The slope on which they form determines whether they are circles, polygons, or stripes.

Pingo

- Dome-shaped mounds formed by the freezing and expansion of groundwater, reaching up to 600 m in diameter and up to 70 m in height.
- Open-system pingos form as groundwater is forced upwards through a crack in the permafrost doming the ground surface.
- A closed-system pingo is formed when a surface lake drains, and the permafrost is advancing underneath. The permafrost forces the water into the centre which then freezes forming the ice core. If the ice core melts, the pingo can collapse.

Loess

Fine sediment that is blown by the high winds in the periglacial environment, sometimes great distances, and deposited as the wind drops.

Distinctive periglacial landscapes

Arctic Tundra

Location: Northern regions of Canada, Alaska, Russia, and Scandinavia.

Landscapes: Permafrost, pingos, patterned ground, thermokarst lakes (formed when permafrost thaws and creates surface depressions that fill with melted water), and widespread cryosols (permafrost soils, also known as gelisols).

Siberian Yedoma Region

Location: Siberia, Northern Russia.

Landscapes: Extensive permafrost, and gelifluction terraces. Gelifluction is very similar to solifluction. Gelifluction terraces are deposits formed on slopes.

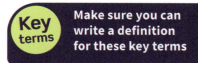 **Make sure you can write a definition for these key terms**

active layer   freeze–thaw weathering   frost heave   ice wedge
loess   nivation   periglaciation   permafrost   pingo   solifluction

Learn the answers to the questions below, then cover the answers with a piece of paper and write as many as you can. Check and repeat.

Questions / Answers

#	Question	Answer
1	What is periglaciation?	The environmental conditions and landforms found at the edges of glaciers, where freezing and thawing processes occur
2	What is permafrost?	A soil or rock that remains frozen throughout the seasons for at least two years
3	What is the active layer in a periglacial environment?	The upper layer of soil that thaws during the summer
4	Give two characteristics of a pingo.	Ice core; covered with a layer of soil or sediment
5	What is frost heave?	The process by which the freezing and expansion of groundwater within soil or rock can uplift the ground surface
6	What is 'patterned ground' in periglaciation?	The organised arrangement of stones and soil, often taking the form of circles, polygons, or stripes
7	How do Milankovitch cycles impact climate on Earth?	Eccentricity, obliquity, and precession contribute to the pacing of glacial-interglacial cycles, shaping the climate variations observed during the Pleistocene
8	When was the Pleistocene epoch?	From approximately 2.6 million to 11,700 years ago
9	When was the Last Glacial Maximum (LGM)?	Around 20,000 years ago
10	What is the cryosphere?	All frozen water on or under the Earth's surface
11	Glacial periods lead to ice advances and interglacial periods cause ice retreats. True or false?	True
12	Name four places in the UK that show Pleistocene glaciation relict landforms.	Four from: Lake District / Scottish Highlands / Snowdonia / Brecon Beacons / Cairngorms / East Anglia
13	Name four types of ice mass.	Four from: ice sheets / ice caps / temperate glaciers / polar ice icefields / glaciers
14	Name the four periglacial processes.	Nivation, frost heave, solifluction, freeze–thaw weathering

Put paper here

⚙ KNOWLEDGE

2 Glaciated landscapes and change

2.4 Glacial mass balance

Glacial mass balance system

A glacier is an **open system**, which means it has inputs from and outputs to other systems.

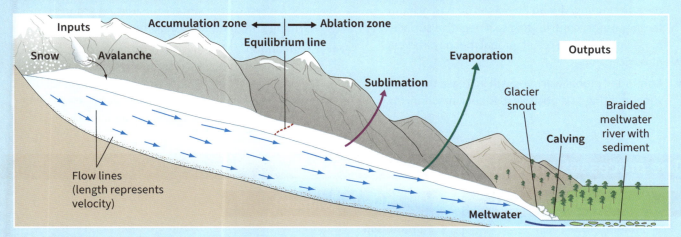

▲ *Figure 1* *Glacial system*

- Mass balance = **accumulation** +/− **ablation**

- The accumulation zone of a glacier is where snowfall exceeds melting. Input occurs through snowfall, wind draft, and avalanching. Over time, this accumulated snow compacts to form glacial ice.

- The equilibrium line marks the zone where accumulation is balanced by ablation over a 1-year period.

- The ablation zone of a glacier is the other side of the equilibrium line and is where outputs from the glacier (melting, sublimation, calving, and avalanches) exceed inputs. Over time, the glacier will reduce in size.

- Feedback loops operate within glacial systems, which are open and interact with other systems such as the hydrosphere and lithosphere.

Importance of positive and negative feedback

An open system has inputs and outputs from exogenous sources (outside the system), e.g. snow from the atmosphere.

Feedback loops operate within the glacial systems, which are open and interact with other systems such as the hydrosphere and lithosphere.

- **Positive feedback**: Processes within the system amplify the inputs to lead to rapid change. In a glacial system this can mean either increased accumulation (and growth and advance of the glacier) or it can mean increased ablation, and therefore a retreat and decrease in glacier size.

- **Negative feedback**: Processes balance any changes in the inputs to return the system to equilibrium. Under this feedback loop, the glacier size and extent remains the same.

2 Glaciated landscapes and change

2.4 Glacial mass balance

Greenland ice sheet

- Data shows a mass loss of land ice over recent decades, mainly due to increased air and ocean temperatures, contributing to a negative surface mass balance. As a result, global mean sea levels have risen.

- Continued global warming will increase the rate of ice sheet melting as a positive feedback mechanism.

- Exposed ground is darker than ice and reduces the albedo effect on the surface, increasing ground warming and, therefore, snow melt.

- Greenland's ice sheet is in a 'positive feedback' loop – rising temperatures are increasing melting, and that melting is reducing albedo, which is in turn leading to increasing temperatures. Earth's systems are moving further and further from dynamic equilibrium.

The processes of accumulation and ablation

Accumulation (inputs)
Precipitation (snow)
Avalanches
Wind deposition

Ablation (outputs)
Sublimation
Evaporation
Meltwater
Calving
Avalanches

Reasons for variations in accumulation and ablation

- Glacial **mass balance** refers to the net gain (accumulation) or loss (ablation) of ice and snow on a glacier over a specific period of time and affects whether the glacier grows or shrinks.

- It is calculated using a combination of direct field measurements and remote sensing techniques.

- It is in a constant state of change due to weather and seasonal changes, and climate change.

- Ablation is at its highest in summer, while accumulation is highest in winter.

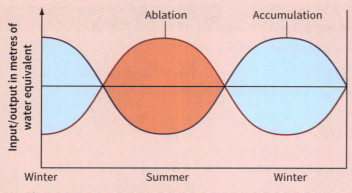

▲ *Figure 2 Seasonal changes in mass balance in a glacier*

> **REVISION TIP**
>
> Make sure you can draw a simple diagram to show glacial mass balance as this will help you to remember this information.

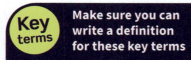 **Key terms** Make sure you can write a definition for these key terms

ablation   accumulation   mass balance   negative feedback
open system   positive feedback   sublimation

2.5 Movement of glaciers

Types of glacier

Different processes explain glacial movement and the variations in rates of movement.

Polar glaciers are cold based and are frozen to the bedrock, causing very slow movement of the glacier downslope.

Temperate glaciers can be warm based in the summer months; where meltwater lubricates the base of the ice, it enables a faster movement downslope.

Important processes in glacial movement

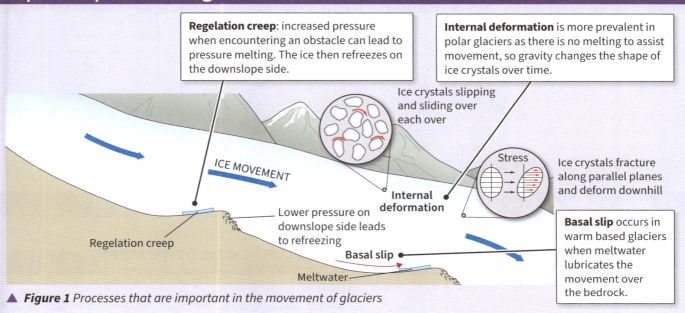

Regelation creep: increased pressure when encountering an obstacle can lead to pressure melting. The ice then refreezes on the downslope side.

Internal deformation is more prevalent in polar glaciers as there is no melting to assist movement, so gravity changes the shape of ice crystals over time.

Ice crystals slipping and sliding over each over

ICE MOVEMENT

Stress

Ice crystals fracture along parallel planes and deform downhill

Regelation creep

Lower pressure on downslope side leads to refreezing

Internal deformation

Basal slip

Meltwater

Basal slip occurs in warm based glaciers when meltwater lubricates the movement over the bedrock.

▲ **Figure 1** *Processes that are important in the movement of glaciers*

Factors controlling the rate of movement

Altitude: height above sea level affects temperature and precipitation.

Basal temperature: basal temperature is a factor that controls basal slip.

Slope: the steeper the gradient of the slope, the faster the glacier will move.

Other factors controlling the rate of movement

Lithology: when the rocks are hard and impermeable, the glacier will move faster as the meltwater won't percolate into the rock, or freeze as easily with it.

Size: the larger the glacier, the greater the mass and pressure causing greater friction at the base.

 Key terms Make sure you can write a definition for these key terms

basal slip   internal deformation   lithology   polar glacier
regelation creep   temperate glacier

 **KNOWLEDGE**

2 Glaciated landscapes and change

2.6 The glacier landform system

Glacial processes

1 Erosion
Erosion by the ice mass, and sediment trapped under and within the ice, takes the form of abrasion and plucking.

2 Entrainment
Rocks from sub-glacial, supraglacial and other sources are incorporated into the ice mass. These range in size from small rock fragments loosened by freeze–thaw all the way up to large boulders.

4 Deposition
Most deposition from glaciers occurs during a loss of energy at the snout, as the glacier retreats, or through meltwater.

3 Transport
Glaciers can carry material long distances. The material can be carried many kilometres within the glacier (englacial), on top of the glacier (supraglacial) or in meltwater at the base of the glacier (sub-glacial).

> **REVISION TIP**
> Cover this page and try to remember the four glacial processes in detail. Check and try again if you make any mistakes.

Development of glacial landforms

Glacial landforms develop at macro-, meso- and micro-scales with distinctive morphologies.

Location	Meaning	Processes and landforms
Sub-glacial	Below or the base of an ice sheet or glacier	Erosion forms striations and roche moutonnée. Eskers can form due to deposition from meltwater.
Marginal	Along the edges and at the end of the glacier	Weathering and deposition create formations such as moraines.
Proglacial	Downslope of the glacier in front of the snout	**Fluvioglacial** processes create outwash plains, meltwater channels and, subsequently, proglacial lakes.
Periglacial	At the margins of the glacier; can be seasonal and extensive	Processes associated with frost, ice, snow and then meltwater; landforms such as pingos, ice wedges.

Landform evidence in upland and lowland areas

The landforms that glacial processes create differ in upland and lowland areas:

- Uplands such as the Lake District are characterised by erosional landforms (e.g. glacial troughs, hanging valleys, corries, tarns).
- Lowland landscapes such as those found in East Anglia are usually characterised by having more depositional landforms (e.g. outwash plains of boulder clay, meltwater channels and lakes).

 Key terms Make sure you can write a definition for these key terms

deposition   entrainment   erosion   fluvioglacial   marginal
periglacial   proglacial   sub-glacial   transport

RETRIEVAL

Learn the answers to the questions below, then cover the answers with a piece of paper and write as many as you can. Check and repeat.

Questions

Answers

	Questions		Answers
1	How are glaciers formed?		From the accumulation and compaction of snow over time
2	What is the glacial mass balance?		The net gain (accumulation) or loss (ablation) of ice and snow on a glacier over a specific period of time
3	Which three processes cause accumulation in the glacier mass balance?	Put paper here	Precipitation (snowfall), wind drift, and avalanching
4	What are the different processes contributing to ablation?		Melting, sublimation, calving, avalanches
5	What factors influence the movement of glaciers?	Put paper here	Altitude, basal temperature, slope, lithology, size
6	Which type of glaciers are cold based and are frozen to the bedrock, causing very slow movement of the glacier downslope?		Polar glaciers
7	On a glacier, what is the equilibrium line?	Put paper here	Marks the zone where accumulation is balanced by ablation over a 1-year period
8	Ablation is at its lowest in summer, while accumulation is highest in winter. True/false?	Put paper here	False
9	What are the two types of glacier?		Polar and temperate
10	Where do marginal glacial landforms form?		Along the edges and at the end of the glacier

Previous questions

Now go back and use these questions to check your knowledge of previous topics.

Questions

Answers

	Questions		Answers
1	What is permafrost?		A soil or rock that remains frozen throughout the seasons for at least two years
2	Give two characteristics of a pingo.	Put paper here	Ice core; covered with a layer of soil or sediment
3	How do Milankovitch cycles impact climate on Earth?		Eccentricity, obliquity, and precession contribute to the pacing of glacial-interglacial cycles, shaping the climate variations observed during the Pleistocene
4	Name four types of ice mass.	Put paper here	Four from: ice sheets / ice caps / temperate glaciers / polar ice / icefields / glaciers
5	Name the four periglacial processes.		Nivation, frost heave, solifluction, freeze–thaw weathering

2 Glaciated landscapes and change

2.7 Glacial erosional processes and landforms

Glacial erosional processes

Process	Description
Abrasion	The grinding action of rock fragments carried by a glacier against the underlying bedrock smooths and polishes the surface.
Plucking or quarrying	Glacier ice freezes to rock, and as the glacier moves, it lifts and plucks off pieces of rock, which are then transported by the glacier.
Crushing	The sheer mass and pressure of a glacier can crush the underlying bedrock, even hard rocks.
Basal melting	Pressure and heat generated by the movement of the overlying ice melts the glacier base. Meltwater acts as a lubricant, facilitating the sliding of the glacier over the bedrock.
Subaerial freeze–thaw	In areas with a climate characterized by freeze–thaw cycles, this mechanical weathering process breaks down the rocks over time.
Mass movement	Downslope movement of rock, soil, and debris under the influence of gravity, including landslides, rockfalls, mudslides, and avalanches.

Erosional landforms

Different erosional landforms are associated with:

- Cirque (or corrie) glaciers – small masses of ice that occupy armchair-shaped hollows in mountains. Often feed into valley glaciers.
- Valley glaciers – larger bodies of ice which move downwards from either a cirque or an ice sheet. Usually follow the courses of former rivers.

Erosional feature	Description	Formation	UK examples
Cirque / corrie / cwm	Bowl-shaped depression in a mountain, often the head of a glacial valley	Glacial erosion deepens and widens cirques, creating these distinctive features.	Red Tarn, Helvellyn, Lake District
Arête	Narrow, sharp-edged ridges	Cirques erode rock on either side, creating a sharp ridge between them	Striding Edge, Lake District
Pyramidal peak	Sharp, pointed mountain peaks	Three or more cirques form back-to-back leaving a sharp peak between them	Helvellyn, Lake District
Glacial trough	U-shaped valleys	Pre-existing V-shaped river valleys are deepened and widened by glaciers, resulting in a U-shaped cross-section	Great Glen, Scotland
Truncated spur / hanging valley	Ridges that were once spurs or ridges between tributary valleys	Truncation occurs through glacial erosion	Honister Pass, Lake District
Ribbon lake	Elongated, finger-shaped lakes	Valleys deepened and widened by glaciers form basins that later fill with water	Lake Windermere, Lake District

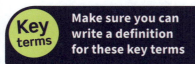 **Key terms** Make sure you can write a definition for these key terms

arête   glacial trough   pyramidal peak
ribbon lake   truncated spur / hanging valley

2.7 Glacial erosional processes and landforms

Formation of a corrie

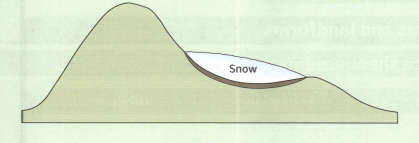

Periglacial conditions
Nivation and frost/ice processes (such as freeze-thaw) slowly increase the size of a depression on the mountainside.

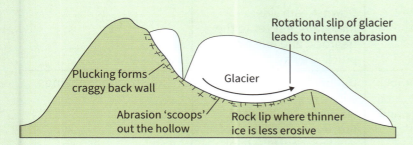

Rotational slip of glacier leads to intense abrasion

Plucking forms craggy back wall

Glacier

Abrasion 'scoops' out the hollow

Rock lip where thinner ice is less erosive

Glacial conditions
Snow turns to ice in the depression. A corrie glacier develops and its mass increases.

Rotational sliding 'scoops out' the hollow by abrasion.

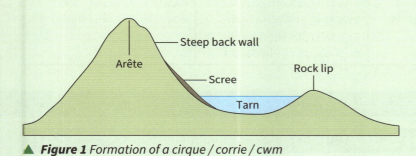

Steep back wall

Arête

Scree

Rock lip

Tarn

Post-glacial conditions
The shape of the corrie is modified by periglacial and then temperate processes.

▲ **Figure 1** Formation of a cirque / corrie / cwm

Formation of arêtes and pyramidal peaks

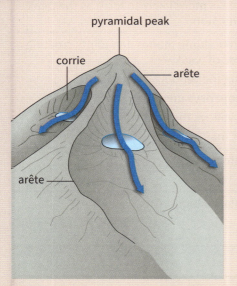

pyramidal peak

corrie

arête

arête

▲ **Figure 2** How arêtes and pyramidal peaks form

⚙ KNOWLEDGE

2 Glaciated landscapes and change

2.7 Glacial erosional processes and landforms

Formation of landforms due to ice sheet scouring

Erosional feature	Description	Formation	UK examples
Roche moutonnée	A smoothly polished and rounded rock surface on the upstream side of the glacier and a rough, irregular surface on the downstream side	Upstream side: abrasion and plucking as regelation slip takes place. Downstream side: meltwater freezes to the bedrock, and loose rock is plucked away.	Honister Pass, Cumbria
Knock and lochan	A small hill or knoll (knock) is associated with a nearby small lake or pond (lochan)	Differential erosion occurs where resistant and weaker rock alternate. Erosion exploits lines of weakness such as joints and minor faults.	In the Scottish Highlands
Crag and tail	Steep, craggy upstream slope (crag) and a gentle, elongated, tapering downstream slope	The crag is formed by plucking and abrasion. The tail is formed by the deposition of glacial till.	Castle Rock (crag) and the Royal Mile (tail), Edinburgh

Influence of differential geology

Type of rock	Examples	Characteristics
Harder, more resistant rocks	**Igneous** and **metamorphic rocks** such as granite, basalt and quartzite	Withstand plucking better More abrasion More vertical valley walls Pyramidal peaks more likely to form as the peak resists total erosion
Softer rocks	**Sedimentary rocks** such as shale, limestone and sandstone	More defined bedding planes and joints creating natural planes of weakness More plucking (depending on climate) Striations more prevalent Widening of valleys through lateral erosion More cirques may form
Alternating rock layers	Softer rocks will erode leaving the harder rocks in place	Roches moutonnées Waterfalls
Permeability and **porosity**	Sedimentary rocks usually have great permeability with more impact from the freeze–thaw cycle	More fluvioglacial landforms

---

Key terms — Make sure you can write a definition for these key terms

crag and tail   igneous rock   knock and lochan
metamorphic rock   permeability   porosity   roche moutonnée
sedimentary rock

2.8 Glacial depositional processes and landforms

Lowland depositional features: till

- **Till** is a mixture of clay, silt, sand, gravel, and boulders that have been eroded, transported and deposited by the glacier. It is angular and poorly sorted.

- Till plains are extensive features left behind as a glacier retreats. The deposition of glacial till over large areas levels out the relief of the underlying land leaving a flat or gently undulating surface.

- Lodgement till is formed when a glacier moves over the landscape, plucking and grinding rocks from the underlying bedrock.

- Ablation till is deposited at the glacier's snout as a result of the melting and ablation of the glacier. It consists of material that was once part of the glacier but has been released as the ice melts. It is the material that composes recessional and terminal moraines.

Ice contact depositional features

Drumlins

Drumlins are elongated hills or mounds of till that are formed beneath a moving glacier and aligned with the direction of glacial movement. They can vary in length from a few hundred metres to several kilometres and be up to around 50 m in height. They usually appear in swarms.

Moraines

Moraines are deposits of poorly sorted, angular sediment (till) found at different locations along the glacier.

Type of moraine	Formation
Lateral moraine	Formed between a glacier and the valley sides
Medial moraine	When two glaciers meet, the lateral moraine from each merges to form a medial moraine
Terminal moraine	Formed furthest downslope and indicates the greatest extent of the glacier
Recessional moraine	Formed at successive points of the retreating snout at each terminal point of recession
Ground moraine	Deposited beneath a glacier as it transports sediment on top of the underlying rock base

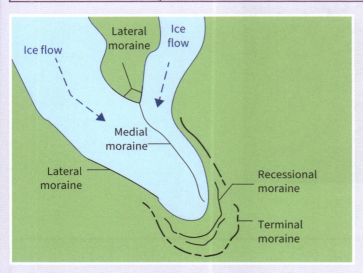

▲ **Figure 1** Plan view of types and locations of moraine

2 Glaciated landscapes and change

2.8 Glacial depositional processes and landforms

Working out previous ice extent, movement and origin

The assemblage of landforms can be used to reconstruct former ice extent, movement and provenance.

Landform	Description	Reconstructing previous extent, movement and origin
Erratics	Large fragments of rock which are clearly different to the geological surroundings, indicating that they have been moved by glaciers.	Once the origin of the rock is established, the direction and distance the glacier moved can be established.
Moraines	Till that remains in various locations after the glacier has melted.	A terminal moraine is found furthest downslope and indicates the greatest extent of the glacier. Recessional moraines are found at successive points of the retreating snout.
Crag and tail	Prominent, steep-sided rock hill (crag) on the up-ice side, followed by a gentle, elongated slope (tail) extending in the down-ice direction. Formed by the erosional action of the glacier on solid rock.	As the glacier flows over the obstacle, it erodes the up-ice side (stoss) through processes like plucking and abrasion. The rock fragments and debris are then deposited on the down-ice side (lee), creating a gently sloping tail e.g. Arthur's Seat in Edinburgh.
Drumlin orientation	Formed of till deposited by the glacier, although a few do have a rock core. The up-glacial side is a gentle slope and there is a steeper tapered slope on the down-ice side.	The alignment of the long edge of a drumlin indicates the direction in which the glacier that formed it was moving. These features are in many parts of northern England, Scotland and other places in northern latitudes.

▲ *Figure 2* Bubble Rock in Acadia National Park, Maine, USA is a glacial erratic

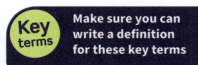
Key terms | **Make sure you can write a definition for these key terms** | drumlin   erratics   ground moraine   lateral moraine   medial moraine   recessional moraine   terminal moraine   till

2.9 Glacial meltwater processes and landforms

Processes of water movement within the glacial system

Meltwater from seasonal melting of warm based glaciers, or glacial retreat is often present in huge quantities which can flow under great pressure and with great velocity.

The processes of water movement in the glacial system are:

- **supraglacial** flow – occurs on the glacier's surface
- **englacial** flow – takes place within the glacier
- **sub-glacial** flow – occurs at the glacier's base.

Glacial and fluvioglacial deposits

Stratification:
Glacial ice carries material of various sizes, and ice flow does not allow for the distinct layering of sediment. As a result, glacial deposits tend to be unstratified or poorly stratified.

Grading:
Grading is when sediment particles are sorted by size, with larger particles at the bottom and finer ones at the top.

Glacial and fluvioglacial deposits

Sorting:
Glaciers transport sediments ranging from clay to boulders. The lack of sorting means that sediments of various sizes are mixed together in glacial deposits.

Glacial deposits are often unstratified, poorly sorted, exhibit imbrication, and lack grading due to the characteristics of glacial ice.

In contrast, fluvioglacial deposits are typically well-stratified, well-sorted, less prone to imbrication, and more likely to display grading, as a result of the processes associated with meltwater streams.

Imbrication:
As a glacier moves, it can pluck and entrain rock fragments from the substrate. These fragments may become imbricated, meaning they are oriented with their long axes pointing in the direction of glacial flow.

2 Glaciated landscapes and change

2.9 Glacial meltwater processes and landforms

Ice contact features

Kames: small, irregular mounds typically found within ice contact margins, such as at the glacier's terminus or along its sides.

Eskers: long, winding ridges of sediment deposited by meltwater streams that form as tunnels or channels within or beneath glaciers. Sediment is deposited as the glacier melts.

Kame terraces: flat-topped hills composed of sediment deposited by meltwater streams.

▲ **Figure 1** *Esker in Manitoba, Canada*

Proglacial features

Landforms and geological structures found in the vicinity of a glacier, often associated with its snout or the area immediately downslope of it.

Sandurs (outwash plains): extensive, flat areas of sediment deposited by meltwater streams as they flow away from a glacier. They are important for agriculture and building materials.

Meltwater channels: a steep valley eroded by meltwater into bedrock that then transports sediment.

Proglacial features

Proglacial lakes: form in front of glaciers, often dammed by moraines. They can grow and retreat, which influences the development of fluvioglacial landforms and can lead to glacial outburst floods.

Kettle holes: large blocks of ice are left behind on glacial retreat. When these melt, large depressions can be left, which may then fill with water.

> **REVISION TIP**
>
> Make a blank table of all the features you need to revise under different categories, then fill it in with definitions and explanations before checking if you are right.

 Key terms — Make sure you can write a definition for these key terms

englacial   esker   grading   imbrication   kame   kettle holes
proglacial lakes   sandurs   stratification   sub-glacial   supraglacial

Learn the answers to the questions below, then cover the answers with a piece of paper and write as many as you can. Check and repeat.

Questions | Answers

	Questions	Answers
1	Name three processes of water movement in the glacial system.	Supraglacial flow, englacial flow, sub-glacial flow
2	What are drumlins?	Elongated, streamlined hill composed of glacial sediment
3	What are erratics?	Large fragments of rock which are clearly different to the geological surroundings, indicating that they have been moved by glaciers
4	What are kames?	Small, irregular mounds typically found within ice contact margins
5	Why are outwash plains (sandurs) important in fluvioglacial environments?	For agriculture and building materials
6	What are kame terraces?	Flat-topped hills composed of sediment deposited by meltwater streams
7	How are eskers formed?	As the glacier melts, water carrying sediment flows through tunnels in the glacier, depositing material on the glacier bed
8	What is a meltwater channel?	A steep valley eroded by meltwater into bedrock that then transports sediment
9	What is till?	A mixture of clay, silt, sand, gravel, and boulders that have been eroded, transported and deposited by a glacier
10	Where does supraglacial flow occur?	On the glacier's surface

Put paper here

Previous questions

Now go back and use these questions to check your knowledge of previous topics.

Questions | Answers

	Questions	Answers
1	Which processes cause accumulation in the glacier mass balance?	Precipitation (snowfall), wind drift, and avalanching
2	What factors influence the movement of glaciers?	Altitude, basal temperature, slope, lithology, size
3	Which type of glaciers are cold based and are frozen to the bedrock, causing very slow movement of the glacier downslope?	Polar glaciers
4	How are glaciers formed?	From the accumulation and compaction of snow over time
5	Where do marginal glacial landforms form?	Along the edges and at the end of the glacier

Put paper here

2 Glaciated landscapes and change

2.10 Value of glaciated landscapes

Environmental and cultural value

Relicts and active glaciated landscapes have environmental and cultural value.

Polar scientific research	Wilderness recreation	Spiritual/religious associations
Antarctica's vast ice sheets and glaciers provide critical data for understanding climate change, ice dynamics, and past climatic conditions. There are up to 10,000 scientists working at 59 research bases in Antarctica. Scientific research also takes place in the Arctic Ocean and on the land.	The Swiss Alps exemplify wilderness recreation in glaciated areas. Tourists visit areas like the Jungfrau region for activities such as hiking, mountaineering, and skiing, appreciating the alpine landscapes, glaciers, and pristine wilderness.	Many high-altitude cold environments hold special associations. Mount Kailash in Tibet is considered sacred in Hinduism, Buddhism, Jainism, and Bon. Pilgrims undertake the challenging *kora* (circumambulation) of the mountain, integrating cultural and spiritual values. In the Arctic, there are over 40 different indigenous ethnic groups with many languages and cultural traditions.

Economic value

Glaciated landscapes are also important economically.

	Economic importance	Examples
Farming	Many hill farms, where sheep and other livestock can be pastured, exist on relict glaciated areas. Heated greenhouses have allowed crops to be grown even in cold climates, allowing for greater local food security.	Sheep farming for meat and wool in Lake District, UK, and in much of New Zealand.
Mining	It is estimated that 30% of undiscovered gas and 13% of unexploited oil deposits, worth $trillions, are in the Arctic. As the Arctic melts, the area will become more accessible, and more mineral exploration will take place.	Lomonosov Ridge is disputed between Russia and other Arctic nations, and is very rich in oil and natural gas.
Hydroelectric power	Downslope of glaciers, huge volumes of meltwater can be stored in deep valley reservoirs, and then height differences utilised to produce hydro-electric energy.	Diablo Dam in the North Cascades of Washington State, USA generates 689,400,000 KWh p/a for the Seattle area.
Tourism	Cold environments have opened up to tourism, thanks to technical clothing and modern technology to ensure safety. Conserved relict environments can be very beautiful, attracting many hikers and other outdoor sportspeople.	Lake District, Alps, Mt Everest Skiing Hiking Mountaineering
Forestry	**Boreal forests (taiga)** contain about 30% of the world's trees, and are extensively used for forestry of especially pine and larch.	Countries and regions such as Russia, Canada, Scandinavia and Alaska produce millions of tonnes of timber each year.

Environmental value

Glaciated landscapes are unique and of major environmental value because they play an important role in the maintenance of natural systems, including both the water and carbon cycles.

- Glaciers are essential indicators of climate change, helping scientists monitor environmental shifts.
- They also act as water reservoirs, supplying freshwater to downstream ecosystems and human communities.
- **Tundra** and high-altitude areas have their own distinct ecosystems.

- These ecosystems contribute significantly to global **biodiversity** as they provide unique habitats.
- Tundra ecosystems are fragile and sensitive to change. Once damaged, they may never recover.

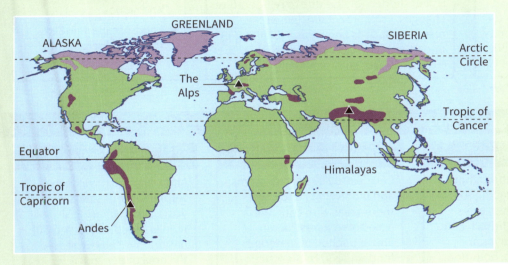

◀ **Figure 1** Tundra systems occur in the Arctic and on mountain slopes close to the snowline – here, the climate is cold, rainfall is low and the land is snow-covered for most of the year

Maintaining natural life support systems

- Boreal forests make up the largest terrestrial (land-based) ecosystem.
- Arctic and highland forests, peatlands and very slowly decaying plant matter are a massive **carbon sink**, the largest on land.
- Soils store more carbon than the trees, although with increased temperatures, this difference could be declining rapidly.
- Massive wildfires have been burning these forests (e.g. 12,000 km² in Alaska in 2022) contributing more greenhouse gases to the atmosphere.
- The cryosphere is an integral part of the hydrological cycle providing a store of 67% of global freshwater. As glaciers melt seasonally, they provide water downstream. However, climate change impacts these stores and they are contributing to sea level rise.
- As ice cover is reduced pale surfaces are replaced by dark ones, less solar radiation is reflected back into space and the **albedo** of the Earth falls.

REVISION TIP

Remember to revise your understanding of the enhanced greenhouse effect and the role of the cryosphere.

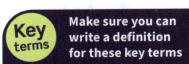 **Key terms** Make sure you can write a definition for these key terms

albedo   biodiversity   boreal forest (taiga)   carbon sink   tundra

⚙ KNOWLEDGE

2 Glaciated landscapes and change

2.11 Threats to glaciated landscapes

Threats from natural hazards

Avalanches

Occur when the mass (downward force) of snow and ice is greater than the opposing force of friction, which can also be enhanced due to melting. Velocity can be as much as 200 km/hr, but is generally less than 60 km/hr.

Glacial outburst floods (GLOFs)

Occur when meltwater accumulates behind ice barriers or moraine, creating a risk of sudden release due to temperature changes, volcanic activity (when the resulting **GLOF** is known as a jökulhlaup) or earthquakes. GLOFs can pose threats to downstream communities.

Preventative measures

- Early warning systems
- Monitoring glacial lakes and slopes
- Land-use planning to reduce vulnerability

Threats from human activities

Leisure and tourism

Huge increase in cruise ships throughout the Arctic and to Antarctica, leading to sewage and blast discharged into the seas, altering pH /salinity and introducing invasive species. In the Himalayas and Alps, footpath and soil erosion, pollution, and demand for resources are all increased. Increasing global population and accessibility of distant locations has led to an exponential growth in tourism in even extreme environments.

Reservoir construction

Can destroy local ecosystems and biodiversity and cover more areas with concrete. Associated infrastructure and extraction of materials for dams add to the issues. Deeper reservoirs such as those in the Alps contain oxygen-poor water, causing further problems downstream when water is released.

Urbanisation

Alpine valleys have been settled for thousands of years, but increasing size of homes (doubled since 1950) has dramatically increased the area under concrete and reduced biodiversity. Increased building of tourist resorts has also contributed. No international standards.

Landscape degradation

Human activity can degrade the landscape and fragile ecology of glaciated landscapes, and reduce its resilience over time.

> **REVISION TIP** 📝
>
> Remember to think about evaluating impacts – which have the most significant effects?

Soil erosion: leads to reduced soil fertility and an increased risk of landslides and sediment entering water bodies. It can also expose underlying bedrock, which disrupts the natural landscape. Also leads to massively reduced carbon sequestration.

Landslides: glaciated areas often have steep terrain and loose, unconsolidated glacial deposits. Human activities that disturb the slopes can trigger landslides, which can damage ecosystems, bury vegetation, and disrupt watercourses, potentially leading to downstream flooding.

Landscape degradation

Trampling: vegetation cover is essential for stabilising soil and preventing erosion. Trampling can disrupt this vegetation, leading to soil erosion and changes in the fragile ecosystems that are adapted to the harsh glacial conditions.

Deforestation: deforestation can increase the risk of soil erosion, alter water regimes, and lead to the loss of habitat for species adapted to these environments.

Climate change

Climate change is having a major impact on glacial mass balances. This may lead to significant disruption of the hydrological cycle.

Meltwater

Warming temperatures contribute to surges in runoff and glacier melt, affecting downstream water availability. While this may initially lead to increased river discharge, it will eventually result in long-term reductions. Sublimation is also increasing. The risk of GLOFs increases as glacial lakes grow in size due to accelerated melting.

River discharge

Over time, reduced glacier mass and altered precipitation patterns lead to a decline in river discharge, particularly during dry seasons, impacting water availability for downstream ecosystems and human communities.

Sediment yield

Changes in glacial meltwater patterns can alter sediment erosion and transport, affecting sediment yield to downstream areas. This can impact river morphology, sediment deposition in floodplains, and the overall sediment dynamics of river systems and the ecosystems that rely on them.

Water quality

Increased sediment transport, altered nutrient levels, and changes in water temperature can impact aquatic ecosystems and have implications for water quality. Additionally, glacial meltwater may release previously stored pollutants or contaminants into downstream water bodies.

Glacial retreat and changes in ice volume are indicators of climate change.

> **REVISION TIP**
>
> Make sure you know some facts and figures about your case studies – be specific.

> **REVISION TIP**
>
> What do you think is the biggest climate change issue? Forming an opinion on this will help you to remember the facts.

▲ **Figure 1** *Eqi Glacier in Greenland has retreated several kilometres over the last century and climate change is only making this worse*

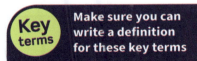
Key terms — Make sure you can write a definition for these key terms

deforestation   glacial outburst flood (GLOF)   landslide
soil erosion   trampling

2 Glaciated landscapes and change

2.12 Managing glaciated landscapes

Different stakeholders

Different stakeholders are involved, using a range of approaches.

Global organisations

UNESCO, other UN bodies and signatories to international agreements such as: the United Nations Convention on the Law of the Seas (UNCLOS), the Antarctic Treaty and the Alpine Convention.

Local, regional and national government agencies

Such as National Parks and Wildlife Reserves.

NGOs (non-governmental organisations)

Operational charities such as Greenpeace and advocates who lobby for more environmental protections such as WWF. NGOs can also include local residents who pressure governments to act.

Other stakeholders

Anyone else who has an interest in the area, including local business owners, landowners, tourists, etc.

Different approaches

Protection

- Conservation areas; scientific research areas; special ecosystem preservation orders

Sustainable management

Balancing conservation with human activities:

- Implementing responsible eco-tourism
- Controlling urbanisation and industrial development
- Reducing deforestation
- Promoting renewable energy sources
- Collaboration among stakeholders, including local communities, scientists, and policymakers

Multiple economic use

Reducing risk from individual activities by diversifying the local economy through:

- Hydropower, wind and solar energy

- Carbon sequestration projects
- Nature-based tourism – e.g. bird-watching
- Environmental education programs
- Traditional handicrafts and artisanal products

Legislative frameworks

- Water and energy conservation
- Climate and transport initiatives
- Zoning regulations and environmental laws

Exploitation

- Unregulated resource extraction leads to further environmental crisis
- International collaboration is needed to prevent exploitation

Climate change

Climate change means successful management of glacial landscapes is increasingly challenging. There is a need for coordinated approaches at global, national and local scale. The future is uncertain, with climate change the greatest risk, and mitigation and adaptation is essential. Glacial retreat and changes in ice volume are indicators of climate change.

 Key terms Make sure you can write a definition for these key terms

multiple economic use   sustainable management

⇄ RETRIEVAL

Learn the answers to the questions below, then cover the answers with a piece of paper and write as many as you can. Check and repeat.

Questions

1. List three threats to glaciated areas from human activities.
2. List three ways in which glaciated landscapes are economically important.
3. How does glacial retreat impact downstream communities?
4. How does climate change affect albedo in glaciated areas?
5. What changes in glacial areas can act as climate change indicators?
6. Give one strategy for the sustainable management of glaciated areas.
7. How do glaciers contribute to rising sea levels?
8. Give three different approaches to managing glaciated landscapes.

Answers

Put paper here

1. Urbanisation, leisure and tourism, reservoir construction
2. Three from: farming / mining / hydroelectric power / tourism / forestry
3. Alters water availability and increases the risk of glacial lake outburst floods (GLOFs)
4. As ice cover is reduced, less solar radiation is reflected back into space and the albedo of the Earth falls
5. Glacial retreat and changes in ice volume
6. One from: implementing responsible eco-tourism / controlling urbanisation and industrial development / reducing deforestation / promoting renewable energy sources / collaboration among stakeholders
7. Released meltwater is eventually discharged to the oceans
8. Three from: protection / sustainable management / multiple economic use / legislative frameworks / exploitation

Previous questions

Now go back and use these questions to check your knowledge of previous topics.

Questions

1. What is periglaciation?
2. On a glacier, what is the equilibrium line?
3. What are drumlins?
4. Why are outwash plains (sandurs) important in fluvio-glacial environments?
5. What are kames?

Answers

Put paper here

1. The environmental conditions and landforms found at the edges of glaciers, where freezing and thawing processes occur
2. Marks the zone where accumulation is balanced by ablation over a one-year period
3. Elongated, streamlined hills composed of glacial sediment
4. For agriculture and building materials
5. Mounds of sorted sediment deposited by meltwater in openings on the glacier's surface

Exam-style questions

1 Study **Figure 1**. Explain how periglacial processes have contributed to the development of this landscape. **(6)**

▲ *Figure 1 Icelandic highlands*

2 Suggest an explanation for the data shown in **Figure 2**. **(6)**

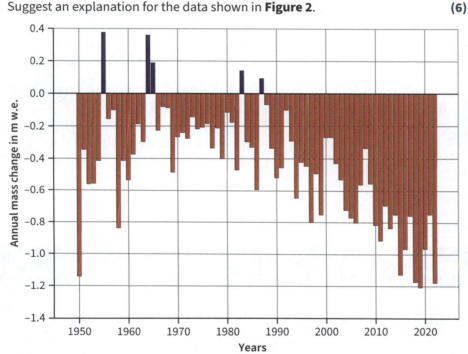

▲ *Figure 2 Annual mass balance of global reference glaciers with more than 30 years of ongoing glaciological measurements. Annual mass change values are given on the y-axis in metre water equivalent (m w.e.) which corresponds to tonnes per square metre (1,000 kg m²).*

3 Using **Figure 3** and your own knowledge, explain the social and economic impacts of a glacial outburst flood. **(6)**

◀ *Figure 3 Boulders of ice on road due to a glacial outburst (jökulhlaup) in Iceland*

4   Explain how moraines can be used to reconstruct former ice extent.   (6)

5   Explain the data in **Figure 4** in relation to global changes in climate.   (6)

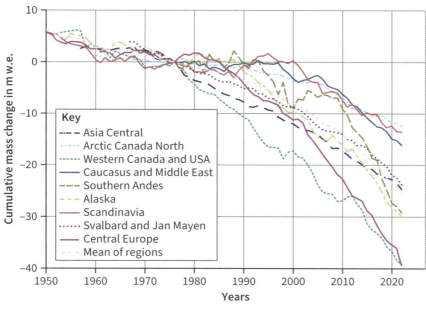

▲ **Figure 4** *Regional and global mean cumulative glacier mass change relative to 1976 based on data from reference glaciers. Cumulative values are given on the y-axis in metre water equivalent (m w.e.) which corresponds to tonnes per square metre (1,000 kg m²).*

6   Study **Figure 5**. Explain the contribution of periglacial processes to the development of this landscape.   (6)

EXAM TIP

For the AO2 marks, state evidence that can be seen in the Figure to make clear connections between the figure and the question.

▲ **Figure 5** *An aerial view of Arctic tundra in autumn: Northwest Territories, Canada*

Exam-style questions

7   Explain how periglacial landforms can be regarded as distinctive.   (8)

8   Explain the role of moraines in providing evidence for past glacial activity.   (8)

9   Explain how periglacial areas can be used to study climate change.   (8)

EXAM TIP

All 8 marks in these 8-mark questions are for AO1. Make sure you show detailed knowledge and understanding of the processes and concepts in the question.

10   Explain how understanding the glacial mass balance contributes to mitigating climate change impacts on glacial processes.   (8)

11   Explain how past, current and future climate changes impact formation of periglacial landforms.   (8)

12   Explain the role of fluvioglacial processes in the formation of landforms within glaciated regions, using named examples.   (8)

13   Explain the relationship between fluvioglacial landforms and hydrological systems in glaciated regions, and their impact on water resource management.   (8)

14   Explain how fluvioglacial landforms are formed and describe their distinct characteristics compared to typical glacial landforms.   (8)

15   Explain the significance of fluvioglacial landforms in understanding past glacial activity and environmental changes.   (8)

16   Evaluate the significance of glacial meltwater in creating distinctive glaciated landscapes.   (20)

17   Evaluate the view that opportunities for economic development in glaciated environments far outweigh any environmental concerns.   (20)

18   Evaluate the relative importance of erosional and depositional processes in creating distinctive fluvioglacial landscapes.   (20)

EXAM TIP

Evaluating a view involves reviewing both sides of the argument and bringing it together to form a conclusion. Points should analyse evidence such as strengths, weaknesses, alternatives, and relevant examples.

19   Evaluate the view that strategies to manage a glacial environment in an upland area (either active or relict) need to have stakeholder involvement.   (20)

20   Examine the role of glacial erosion in creating distinctive landforms in upland areas using named examples.   (20)

21   Evaluate the view that strategies implemented to manage glacial hazards, such as avalanches and glacial lake outburst floods, in upland environments, are rarely effective.   (20)

 **KNOWLEDGE**

3 Coastal landscapes and change

3.1 The coast and wider littoral zone

The littoral zone

The **littoral zone** is the part of the coast that is affected by the action of the waves, and so is constantly changing.

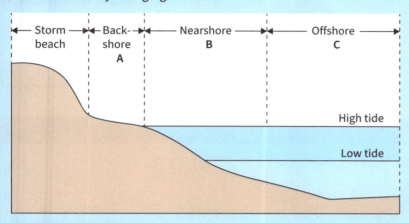

A Backshore: the area of the beach between the high tide line and the furthest extent of the beach.

B Nearshore: the area of the beach closest to the sea where waves break and swash runs up the beach.

C Offshore: the area extending out from the point at which the waves break into the sea into deeper water.

▲ **Figure 1** *The littoral zone*

Classifying coasts

There are different ways to classify areas of coast.

Long-term criteria	Short-term criteria
Geology – coasts can be classified based on their rock type and structure: • Concordant coastlines are those where different rock types are parallel to the shoreline. • Discordant coastlines are those where different rock types are perpendicular to the shoreline. • Rocky coasts are formed by more resistant rocks as they can withstand physical processes. • Coastal plains are formed in areas of low relief as a result of deposition of sediment. Changes of sea level can create distinctive coastlines: • Rises in sea level can create submergent coastlines with landforms such as rias (submerged river valleys) and fjords (submerged glacial valleys). • A fall in sea level can create emergent coastlines where features such as relict cliffs and raised beaches can be found.	Waves can influence the processes that take place and landforms that are formed: • A high-energy environment is characterised by powerful, destructive waves and high rates of erosion. • A low-energy environment has a higher rate of deposition due to less powerful, constructive waves for most of the year, characterised by sandy and estuarine coasts. Tidal range influences the area which is affected by wave action: • Some coastlines have small tidal ranges (e.g. enclosed coasts like in the Baltic Sea). • Some have much larger tidal ranges (e.g. the Severn Estuary in the UK). Rivers are a main source of sediment input for the coastline: • Rivers with large loads can create different types of deltas. • Periodic increases in sediment discharge, such as during storms, can create depositional features such as bars.

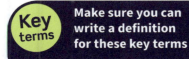 **Key terms** Make sure you can write a definition for these key terms

geology   littoral zone

3 Coastal landscapes and change

3.2 Geology and other factors in the development of coastal landscapes

Geology and coastal morphology

Geology has two aspects:

1. **Lithology**: the geological structure of a rock

2. **Rock type**: resistance, permeability.

These two aspects affect coastal **morphology** (the shape and structure of coastal landscapes), coastal features, such as the type of coastline, and cliff profiles.

> **REVISION TIP**
>
> Remember that lithology is the study of the general physical characteristics of rocks, whereas morphology is the study of the geological structure, shape, or form of a feature.

Concordant and discordant coastlines

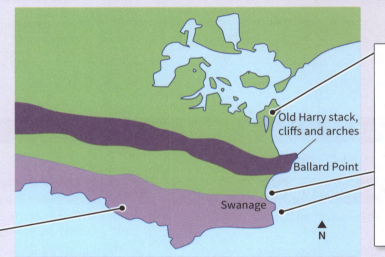

Concordant coastlines: different rock types run parallel to the coastline.

The **Dalmatian** type (see 3.6) has long islands and coastal inlets that lie parallel to the coastline. They are commonly high-energy, submergent coastlines. The **Haff** type has lagoons created by spits that form parallel to the coast.

Old Harry stack, cliffs and arches

Ballard Point

Swanage

N

Discordant coastlines: different rock types run *perpendicular* to the coastline. Less resistant rock **strata** are eroded to create bays and more resistant rock strata resists erosion to form headlands either side (differential erosion).

▲ *Figure 1 Concordant and discordant coastlines*

Folding

Layers within rock can become folded due to tectonic movement.

In anticlines, the folds move upwards so the oldest rocks are in the centre.

In synclines, the rocks fold downwards so the youngest rocks are in the centre.

▲ *Figure 2 Folded limestone on Crete, Greece*

Cliff profiles

Joints and faults are points of weakness at which rocks are vulnerable to weathering and erosional processes. They can be exploited to form wave-cut notches and caves in a cliff.

The dip of a rock describes the angle between the horizontal and the rock strata.

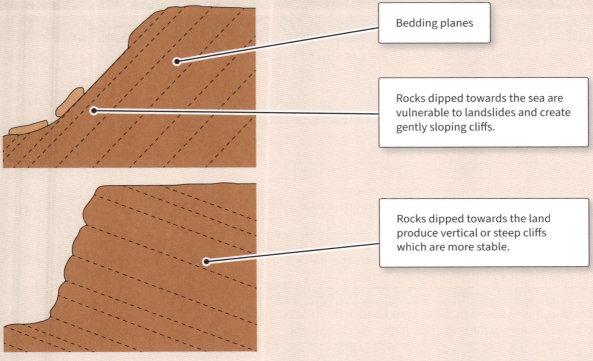

Bedding planes

Rocks dipped towards the sea are vulnerable to landslides and create gently sloping cliffs.

Rocks dipped towards the land produce vertical or steep cliffs which are more stable.

▲ **Figure 3** *The dip of rock at a coastline*

Layers of less resistant strata within the rock erode more easily into wave-cut notches. Permeable rocks are less resistant to weathering processes. A permeable layer of rock above an impermeable layer will become saturated with water and more susceptible to mass movement.

Bedrock lithology

The type of bedrock can influence the rate of coastal recession.

Bedrock type	Formation and characteristics	Effect on rate of coastal recession
Igneous (e.g. basalt and granite)	• Formed by molten rock which has cooled and solidified either on Earth's surface (extrusive) or underground (intrusive) • Impermeable with few faults	• These rocks are some of the most resistant to marine erosion and weathering, so leads to slow rates of coastal recession
Sedimentary (e.g. limestone and sandstone)	• Formed from sediments which have become compacted to form rocks with a layered structure • Permeable with many bedding planes	• Susceptible to chemical weathering (e.g. carbonation) • Less resistant to marine erosion, so leads to fast rates of coastal recession
Metamorphic (e.g. slate and gneiss)	• Formed when rocks are subject to heat and pressure changing them into different rocks with crystalline structures • Impermeable	• More resistant to marine erosion and weathering, so leads to slow rates of coastal recession
Unconsolidated (e.g. boulder clay)	• A mixture of rock material formed by the deposition of sediment carried by glaciers (till)	• Very easily eroded and weathered, so leads to fast rates of coastal recession

3 Coastal landscapes and change

3.2 Geology and other factors in the development of coastal landscapes

Vegetation

The roots of vegetation growing on sand dunes and in salt marshes trap sediments and stabilise the structure of the dune or marsh.

Sand dune succession

> Sand dunes begin to form when sand accumulates around an obstruction on a beach, such as a strand line or fence.

> The first plants to grow on embryo dunes are pioneer colonising species, such as couch or lime grass. They have long roots, which stabilise the sand.

> The foredune is stabilised further by **xerophytes** (plants that can survive with little water) such as marram grass.

> The fixed and mature dunes have a higher humus content, formed as plants die and break down into the soil, and a larger variety of plants.

◀ **Figure 4** *Marram grass can help to stabilise sand dunes*

Salt marsh succession

> Salt marshes are formed in sheltered areas of coastline such as estuaries and bays.

> The first plants to grow in salt marshes are small pioneer plants and algae. Both help to bind the mud and clays together and trap sediment.

> **Halophytes** (salt-tolerant plants) establish themselves next and decompose to add organic matter to the soil.

> Over time, a wider variety of plants will be able to survive in the salt marsh.

REVISION TIP

Mnemonics can help you recall the sequence of events of sand dune and salt marsh succession. Identify the key words for each and write them out. Create a memorable sentence where the initial letters are the same as those of the words in the sequence.

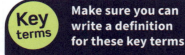 **Key terms**

Make sure you can write a definition for these key terms

Dalmatian   Haff   halophytes   lithology   morphology   strata   xerophytes

RETRIEVAL

Learn the answers to the questions below, then cover the answers column with a piece of paper and write as many as you can. Check and repeat.

Questions

		Answers
1	Define the littoral zone.	The part of the coast that is affected by the action of the waves
2	What is the backshore?	The area of beach between the high tide line and the furthest extent of the beach
3	What is the nearshore?	The area of beach closest to the sea where waves break
4	What is the offshore?	The area extending out from the point at which the waves break into the sea
5	What two long-term criteria can be used to classify coasts?	Geology and changes of sea level
6	What three short-term criteria can be used to classify coasts?	Waves, tidal range, rivers
7	Define lithology.	The geological structure of a rock
8	Define coastal morphology.	The shape and structure of coastal landscapes and features
9	Describe the geology of discordant coastlines.	Different rock types run perpendicular to the sea
10	Which type of coastline has different rock types that run parallel to the sea?	Concordant coastline
11	Which type of concordant coastline has long islands and coastal inlets that lie parallel to the coastline?	Dalmatian
12	Which type of concordant coastline has lagoons created by spits that form parallel to the coast?	Haff
13	What is dip?	The angle between the horizontal and rock strata
14	Name two bedrock types that are more resistant to erosion.	Igneous and metamorphic
15	What are xerophytes?	Plants that can survive with little water
16	What are halophytes?	Salt-tolerant plants

Put paper here

3 Coastal landscapes and change

3.3 Coastal erosion processes and landforms

Waves

Waves are formed when wind blows over the surface of the sea. The type and size of wave are influenced by wind speed, duration, strength, and the fetch.

▲ **Figure 1** *Destructive waves (high energy)*

REVISION TIP

Learn some specific similarities and differences between constructive and destructive waves so you can give detail in your exam answers.

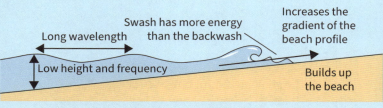

▲ **Figure 2** *Constructive waves (low energy)*

REVISION TIP

Using diagrams to help you revise will enable you to picture the key features and processes of coastal landscapes more easily.

Temporal changes to beach morphology

The shape of beaches changes over different time scales.

- Seasonal: destructive waves are more common in the winter; constructive waves are more common in the summer.

- Daily: features found on beaches reflect the daily changes in swash and backwash. Berms are ridges

of sediment that run parallel along the beach and mark the furthest extent of each high tide.

- During storms: during the highest of tides and during storm events, sediment is pushed to the very back of the beach, creating a storm beach, where it may stay until another storm takes place.

Erosional processes

Hydraulic action: Waves force air into the cracks in cliffs. The repeated increase in pressure forces cracks to widen and pieces break off the cliff.

Corrosion: A chemical process where limestone is dissolved by carbonic acid in the seawater. It is the only erosional process that is not more effective under storm conditions.

The impact of erosional processes is influenced by waves (destructive waves have more erosional power) and lithology (weaker rocks are more susceptible to erosion)

Abrasion: Sediment carried by the waves is thrown at the cliffs and wears away the cliff face.

Attrition: Sediment particles of all different sizes are carried in the waves. They knock against each other to create smoother, smaller, and rounder particles.

Erosional landforms

Erosion can create distinctive landforms: caves, arches, stacks and stumps, cliffs, wave-cut notches and platforms, and headlands and bays (see 3.2).

Caves, arches, stacks, and stumps

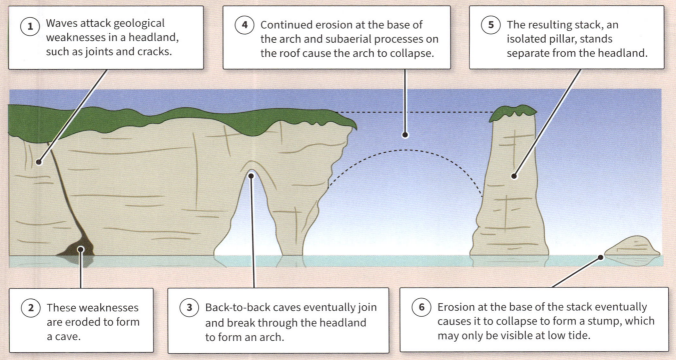

1. Waves attack geological weaknesses in a headland, such as joints and cracks.

4. Continued erosion at the base of the arch and subaerial processes on the roof cause the arch to collapse.

5. The resulting stack, an isolated pillar, stands separate from the headland.

2. These weaknesses are eroded to form a cave.

3. Back-to-back caves eventually join and break through the headland to form an arch.

6. Erosion at the base of the stack eventually causes it to collapse to form a stump, which may only be visible at low tide.

▲ **Figure 3** *Formation of caves, arches, stacks, and stumps at a headland*

Cliffs and wave-cut notches and platforms

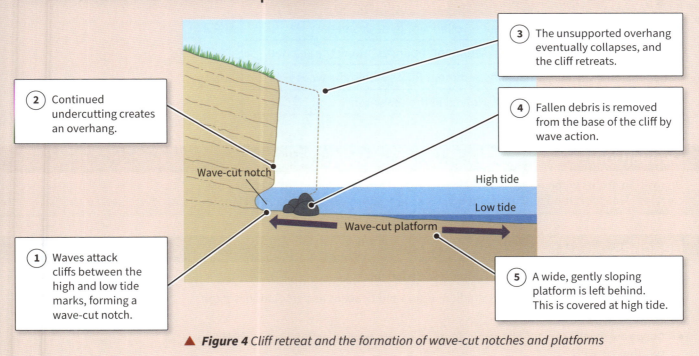

3. The unsupported overhang eventually collapses, and the cliff retreats.

2. Continued undercutting creates an overhang.

4. Fallen debris is removed from the base of the cliff by wave action.

Wave-cut notch

High tide

Low tide

Wave-cut platform

1. Waves attack cliffs between the high and low tide marks, forming a wave-cut notch.

5. A wide, gently sloping platform is left behind. This is covered at high tide.

▲ **Figure 4** *Cliff retreat and the formation of wave-cut notches and platforms*

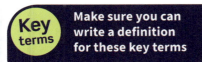

Key terms Make sure you can write a definition for these key terms

abrasion   attrition   corrosion   hydraulic action

3 Coastal landscapes and change

3.4 Transport and deposition

Sediment transportation

Sediment transportation is influenced by the angle at which the wave approaches the coast, the process of longshore drift, and the movement of tides and ocean currents.

Longshore drift

The process of longshore drift transports material along a coastline, in the direction of the prevailing wind.

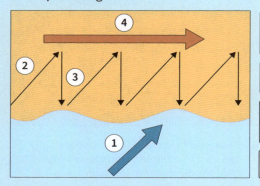

1. Waves approach the coastline at an oblique angle, influenced by the direction of the prevailing wind.

2. Swash pushes material up the beach at this angle.

3. Backwash moves sediment back down the beach, under the influence of gravity, at a right angle to the coastline.

4. The process repeats and sediment is transported along the beach.

▲ **Figure 1** The process of longshore drift

Transportation processes

Process	Explanation
Traction	Large sediment is rolled along the sea bed.
Saltation	Medium-sized sediment 'bounces' along the sea bed.
Suspension	Very small, fine sediment floats within the waves.
Solution	Sediment is dissolved in the seawater.

Deposition

Deposition occurs when waves lose their energy and drop the sediment that they are carrying. Transportation and deposition processes produce distinctive coastal landforms.

Beaches

- Beaches are an accumulation of deposited sediment.
- Shingle beaches have a steeper gradient than sand beaches as percolation is more rapid, leaving less backwash.
- **Cusps** are small crescent-shaped indents which are formed on the beach when two swashes converge.
- An **offshore bar** is a narrow ridge of sediment that runs parallel to the coast. It is formed when backwash removes material from the beach and deposits it in the offshore zone.

Spits

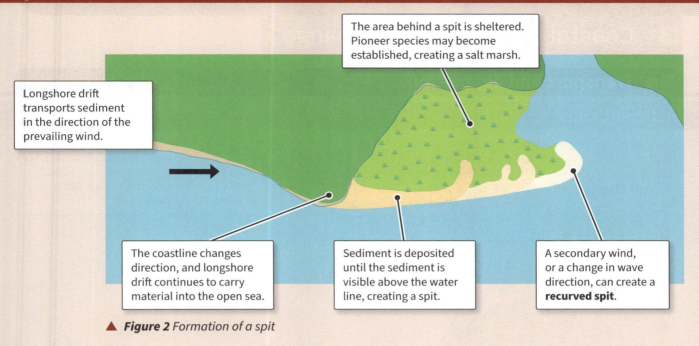

The area behind a spit is sheltered. Pioneer species may become established, creating a salt marsh.

Longshore drift transports sediment in the direction of the prevailing wind.

The coastline changes direction, and longshore drift continues to carry material into the open sea.

Sediment is deposited until the sediment is visible above the water line, creating a spit.

A secondary wind, or a change in wave direction, can create a **recurved spit**.

▲ **Figure 2** *Formation of a spit*

- Spits are more common in areas with low tidal ranges.
- Spits that form across an estuary become cut off when the river current is too strong to allow deposition to continue.
- A spit can join an island to the mainland, creating a **tombolo**.
- If a spit extends across a bay, connecting the headlands, it creates a bar or **barrier beach**, with a lagoon of brackish water forming on the landward side.
- Where there are two opposing wave directions, longshore drift current can converge and **double spits** can form.
- If two double spits join and material is deposited into the sea in a triangular shape, a low-lying headland called a **cuspate foreland** is formed.

The coast as a system

Coasts can be viewed as open systems with inputs, processes, and outputs.

Inputs	Processes	Outputs
Energy from wind and waves	Flows	Removal of sediment to the ocean by currents
Sources of sediment:	Transfer of sediment by:	Evaporation
• eroded sediment	• longshore drift	
• weathering and mass movement	• swash and backwash	
• rivers	Stores of sediment:	
	• depositional landforms (e.g. beaches)	

3 Coastal landscapes and change

3.4 Transport and deposition

Sediment cells

The coastline of England and Wales is divided into 11 stretches of coastline called **sediment cells**. Each cell contains several sub-cells. For example, sediment cell five runs from Portland Bill to Selsey Bill on the south coast and has seven sub-cells.

> **REVISION TIP**
>
> Make sure you know specific details about one sediment cell, such as the cell between Portland Bill and Selsey Bill.

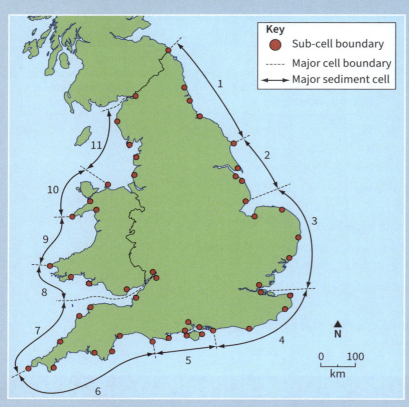

Key
- 🔴 Sub-cell boundary
- ----- Major cell boundary
- ⟷ Major sediment cell

▲ *Figure 3 Sediment cells in England and Wales*

- Sediment cells are closed systems.
- Sediment cells have sources of sediment, transfers of sediment, and sediment sinks (depositional landforms).
- Sediment theoretically stays within each cell, with no transfer between one cell and its neighbour.
- Dynamic equilibrium means that the inputs and outputs are equal, but the system is still in a state of change. Sediment accumulation happens at the same rate as it is removed.

- Positive feedback is when a change in the coastal system leads to a further change. For example, an increase in wave energy will lead to an increase in erosion.
- Negative feedback is when a change in the coastal system leads to a change which returns the system to equilibrium. For example, an increase in deposition will steepen the beach profile, which encourages destructive waves, which will then flatten the beach profile.

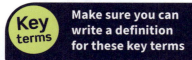 **Key terms** Make sure you can write a definition for these key terms

barrier beach   cuspate foreland   cusps   double spit
offshore bar   recurved spit   sediment cell   tombolo

3.5 Subaerial processes

Weathering

Weathering is a subaerial process which provides a sediment source for the coastal system and influences the morphology of cliffs and the rate of cliff recession.

Chemical	Mechanical	Biological
A process that attacks the minerals within rocks. For example, carbonation occurs when the carbonic acid in rainwater reacts with the calcium carbonate in rocks such as limestone, dissolving the limestone and removing it by solution.	The change in composition of the rock by physical processes, without any chemical change. For example, freeze–thaw weathering occurs when water enters cracks in rocks, freezes and expands, exerting pressure on the rock, and then thaws. This repeated process will eventually lead to rocks splitting apart.	A process where the roots of plants grow into cracks in rocks, widening them and forcing them apart. Burrowing animals can also weather softer rocks and sand dunes.

Mass movement

Mass movement is the downward movement of material under the influence of gravity. It is more likely to occur where cliffs have a weak rock type and lithology.

Blockfall occurs on slopes over 40 degrees. Rock fragments fall from the cliff face and form **talus scree slopes** at the foot of the cliff.

Mass movement

Rotational slumping is more likely to occur in softer rocks, where sediment moves 'en masse' downslope, leaving a **rotational scar** (a curved rupture surface) and a **terraced cliff profile**.

Landslides leave behind a flat rupture surface as the movement of weathered sediment is planar (downslope).

Key terms
Make sure you can write a definition for these key terms

blockfall   mass movement   rotational scar
rotational slump   talus scree slope   terraced cliff profile

Learn the answers to the questions below, then cover the answers column with a piece of paper and write as many as you can. Check and repeat.

Questions

		Answers
1	Give three characteristics of destructive waves.	Three from: greater in height and frequency / short wavelength / backwash has more energy than the swash / remove material from a beach / decrease the gradient of the beach profile
2	What is abrasion?	Sediment carried by waves is thrown at cliffs, wearing away the cliff face
3	Which is the only erosional process that is not more effective in storm conditions?	Corrosion
4	What is longshore drift?	Sediment is transported along a coastline, in the direction of the prevailing wind
5	What is saltation?	Medium-sized sediment bouncing along the sea bed
6	How does a tombolo form?	A spit extends out to reach an island, and joins that island to the mainland
7	When is dynamic equilibrium achieved?	When the inputs and outputs are equal
8	How many sediment cells are there in England and Wales?	11
9	Name a type of mechanical weathering.	Freeze–thaw
10	Name three types of mass movement.	Blockfall, rotational slump, landslide

(Put paper here)

Previous questions

Now go back and use these questions to check your knowledge of previous topics.

Questions

		Answers
1	Describe the geology of discordant coastlines.	Different rock types run perpendicular to the sea
2	Name two bedrock types that are more resistant to erosion.	Igneous and metamorphic
3	Define the littoral zone.	The part of the coast that is affected by the action of the waves
4	What are halophytes?	Salt-tolerant plants
5	Which type of concordant coastline has long islands and coastal inlets that lie parallel to the coastline?	Dalmatian

(Put paper here)

 **KNOWLEDGE**

3 Coastal landscapes and change

3.6 Sea level change

Causes of long-term sea level change

Eustatic sea level change	Isostatic sea level change
A global change in the volume of water in the ocean.	A local change in the land level, leading to a relative change in sea level.

Eustatic sea level change

A global change in the volume of water in the ocean.

- As temperatures increase, the ice stored in ice sheets and ice caps melts, and sea level rises.
- Warmer temperatures also lead to thermal expansion, where the water volume increases, and sea level rises.
- During glacial periods, more water was stored on the land in solid ice form, and sea levels fell.

Isostatic sea level change

A local change in the land level, leading to a relative change in sea level.

- Land can subside due to the weight of the ice stored on it, leading to a relative sea level rise.
- Land once covered in ice is still rising and falling in a process called glacial isostatic adjustment.
- A high rate of deposition may increase the level of the land, leading to a relative sea level fall.
- Tectonic processes can displace the ocean and lead to sea level change.

Effects of long-term sea level change

Emergent coastlines

- Formed when sea levels fall.
- Raised (relict) beaches are found higher than the current sea level.
- Fossil cliffs are often found at the back of raised beaches and were formed by erosional processes before sea levels fell.
- Emergent landforms are affected by subaerial processes (not marine processes).

Submergent coastlines

- Formed when sea levels rise.
- **Rias** are formed when V-shaped river valleys are flooded and submerged.
- **Fjords** are formed when U-shaped glacial valleys are flooded and submerged.
- Dalmatian coastlines (see 3.2) are formed where geological folds run parallel to the sea on a concordant coastline and sea levels rise, flooding the valleys of the folds and leaving narrow islands.

Contemporary sea level change

Contemporary sea level change from global warming or tectonic activity is a risk to some coastlines.

Most scientists believe that anthropogenic global warming is leading to an increase in global mean sea level. It is predicted that the sea level could rise by between 0.43 and 0.84 m by 2100. Low-lying coastal communities, and Small Island Developing States (SIDS), such as the Maldives, are becoming increasingly vulnerable to coastal flooding and erosion.

Coastlines with frequent tectonic activity are also at risk. A large section of eastern Japan lies below sea level. Following the 2011 earthquake, this area became five times larger due to subsidence caused by the tectonic movement.

> **REVISION TIP**
>
> See the Geographical Skills section for help with aerial interpretation of landforms indicating sea level change.

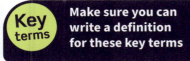 **Key terms** Make sure you can write a definition for these key terms

eustatic   fjord   isostatic   ria

3   Coastal landscapes and change

3.7   Causes of coastal flooding

Coastal recession

```
                    ┌─────────────────────────────────────┐
                    │  Factors affecting coastal recession │
                    └─────────────────────────────────────┘
```

Physical factors that cause recession			Human factors that impact recession	

Geological:
- Softer, weaker rock is more easily eroded.
- Rocks with geological weaknesses (e.g. joints) are more easily exploited by erosional processes.
- Rocks dipped towards the sea are less stable, so more vulnerable to mass movement.

Subaerial:
- Material that is loosened by weathering is more vulnerable to mass movement. For example, blockfalls of material loosened by freeze–thaw can occur, leading to cliff retreat.
- Heavy rainfall can also trigger landslides.

Marine:
- Destructive waves that break at the foot of a cliff have more energy to erode.
- High-energy waves, created in high winds and with long fetches, have more erosive power.
- A narrow beach will be unable to dissipate wave energy as well as a wide flat beach.
- Wave refraction will cause wave energy to be concentrated around headlands and dissipated in bays.

Dredging:
The mining of sediment from the nearshore creates a pit which will be infilled by sediment within the coastal system. This removes more sediment from the beach and can increase rates of erosion.

Coastal management:
Coastal defences, such as sea walls, can deflect or reduce the power of destructive waves and protect the coastline from erosion. Groynes can trap sediment transported by longshore drift. They can also prevent sediment from moving downdrift, leading to sediment starvation, a narrow beach, and increased coastal erosion.

The Nile Delta

The area of the Nile Delta has decreased by almost 3 km² over the last 25 years. The building of the High Aswan Dam has reduced the sediment supply to the delta. Coastal protection, in the form of wave breakers, has accelerated deposition behind them, but increased erosion between them. The area is at increased risk from future sea level rise as the delta subsides, which will flood coastal communities and farmland.

> **REVISION TIP**
>
> Make sure you know about how rapid coastal recession is caused by physical and human factors in one place context.

Factors influencing rates of recession

Rates of **recession** are not constant and are influenced by different factors.

Short-term influences

- Wind direction: wind that is blowing onshore will create waves that attack the coastline.
- Fetch: a long fetch will increase the wave height, shorten the wavelength, and increase the wave energy.
- Weather systems and storms: low pressure weather systems can create storm conditions which increase wind strength and wave energy.

Long-term influences

- Tides: during high tide, the rate of recession is increased as waves can reach the backshore.
- Seasons: winter tends to have more storms which are more likely to increase rates of recession.

Local factors influencing coastal flood risk

Coastal flooding is a significant and increasing risk for some coastlines.

Future sea level rise will exacerbate the risk, particularly for low-lying areas, such as the Maldives, where much of the land is already less than 5 m above sea level and is at risk of becoming uninhabitable in the future.

Removing vegetation and its roots from coastlines can destabilise landforms such as sand dunes, and destroy ecosystems such as mangrove forests. Both of these provide natural protection from flooding.

Local factors increasing the flood risk on some low-lying and estuarine coasts

Places that are higher and well above sea level have a lower risk of coastal flooding.

Human activity, such as the extraction of groundwater, can lead to land subsidence, increasing its vulnerability to flooding.

REVISION TIP

You need to understand how global sea level rise increases the risk of coastal flooding in one place context.

Storm surge events

Storm surges are short-term increases in sea level, which can be caused by **depressions** and tropical cyclones (a rapidly rotating storm system with very low air pressure). The low air pressure 'pulls' the sea level up. Strong winds create large waves and push the water towards the coast. When Hurricane Katrina hit the USA in 2005, the storm surge was over 8 m and overtopped existing coastal protection, flooding 80% of the city of New Orleans.

Climate change

Climate change may increase coastal flood risk, but this is far from certain.

- The effect of climate change on tropical storms is difficult to ascertain due to their rarity and short-lived nature.
- Scientists believe it is likely that the overall frequency of tropical storms will decrease, but the frequency of more intense storms will increase in some areas.
- Scientists are almost certain that mean global sea levels will continue to rise.
- The likely rise by 2100 is uncertain, and depends on the level of future greenhouse gas emissions.

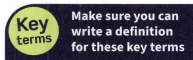 **Key terms** Make sure you can write a definition for these key terms

depression   dredging   recession

RETRIEVAL

Learn the answers to the questions below, then cover the answers with a piece of paper and write as many as you can. Check and repeat.

Questions

1 What is the difference between eustatic and isostatic change?

2 When are emergent coastlines formed?

3 What is a ria?

4 Name the two causes of contemporary sea level change.

5 How can coastal management cause coastal recession?

6 Name the three physical factors that affect the rate of coastal recession.

7 What are the two long-term influences on rates of recession?

8 How can removing vegetation at coasts increase the risk of flooding?

9 What are storm surges?

10 What effects do scientists believe global warming may have on tropical storms?

Answers

Eustatic changes result from a global change in the volume of water in the ocean; isostatic result from a change in the level of the land, with a relative change in sea level

When sea levels fall

A drowned V-shaped river valley

Global warming and tectonic activity

Groynes can stop the movement of sediment along the coast by longshore drift, leading to starvation downdrift

Marine, subaerial, geological

Tides, seasons

Destabilises landforms such as sand dunes, and destroys ecosystems such as mangrove forests, both of which provide natural protection from flooding

Short-term increases in sea level which can be caused by depressions and tropical cyclones

The frequency may decrease, but the frequency of intense storms may increase in some areas

Put paper here

Previous questions

Now go back and use these questions to check your knowledge of previous topics.

Questions

1 What two long-term criteria can be used to classify coasts?

2 What is saltation?

3 Give three characteristics of destructive waves.

4 When is dynamic equilibrium achieved?

5 Name the three types of mass movement.

Answers

Geology and changes of sea level

Medium-sized sediment bouncing along the sea bed

Three from: greater in height and frequency / short wavelength / backwash has more energy than the swash / remove material from a beach / decrease the gradient of the beach profile

When the inputs and outputs are equal

Blockfall, rotational slump, landslide

Put paper here

 KNOWLEDGE

3 Coastal landscapes and change

3.8 Impacts of coastal flooding

Consequences of coastal recession and flooding

The consequences can be significant, especially in areas of dense coastal populations.

Economic losses	Social losses
• Properties and businesses can be completely destroyed by cliff collapse or flooding. • Possessions of residents are lost or damaged. • Homeowners may find themselves in negative equity as properties at increasing risk will lose their value. • Agricultural land can be lost to flooding, or allowed to flood to manage the retreat of the coastline. • Transport infrastructure is at risk: in 2014, storm winds and high sea levels washed away 80 m of train track along the coastal path in Dawlish, UK. • In the UK, 35 power stations, 22 water facilities, and 91 sewage works are at risk from coastal flooding.	• Recreational uses of the coastline are affected when beaches and facilities disappear. • Social and political tensions can arise over how to manage or cope with the risks. • The wellbeing of a community can be affected as people are forced to relocate. • Businesses or agricultural land that is destroyed affects people's livelihoods. • Areas can become less attractive to locals and tourists when infrastructure is at risk or destroyed.

Economic and social consequences of coastal flooding and storm surges

The consequences can be significant, especially in areas of dense coastal populations.

Developing country: Mozambique

In March 2019, Cyclone Idai caused a 4.4 m-high storm surge in the coastal city of Beira in Mozambique. Mozambique is one of the lowest-income countries in the world, with a GDP per capita of US$491.

- The storm surge caused flooding up to 6 m in depth.
- There were over 600 deaths, and more than 111,000 properties and 700,000 hectares of crops were destroyed.
- There were outbreaks of waterborne diseases, such as cholera.
- The cost of the damage and disruption to economic activities was estimated at US$3 billion.

Developed country: UK

In December 2013, a low-pressure system, high tides, and strong winds created the largest storm surge to hit the east coast of the UK in 60 years.

- Homes collapsed into the sea as high tides eroded the cliff beneath.
- 10,000 homes were evacuated.
- There were three injuries.
- 720 residential and commercial properties were flooded.
- The total economic cost to England and Wales from flooding between December 2013 and March 2014 was estimated at £1000 million to £1500 million.

 REVISION TIP

Make sure you revise specific examples of the impacts of coastal flooding, e.g. statistics.

 REVISION TIP

Learning key facts and statistics will help you discuss and compare them in detail in an exam. But don't just list facts; use them to develop your explanation.

Environmental refugees

Environmental refugees will be forced to leave their homes as the effects of climate change put more coastal communities at risk from flooding and storm surges. For example, sea level rise could cause Bangladesh to lose 17% of its land by 2050, creating over 20 million refugees.

3 Coastal landscapes and change

3.9 Hard and soft engineering

Hard engineering

Some people argue that **hard engineering** can negatively affect neighbouring coastlines and limit the ability of the coastline to respond naturally to changes in the coastal system.

Hard engineering approaches directly affect the physical coastal processes. They are expensive, but effective at preventing coastal recession and protecting against flooding

Groynes: structures built at right-angles to the coast that stop longshore drift, trap sediment, and build up a beach. BUT they can lead to sediment starvation in other areas.

Sea walls: can be recurved and repel wave energy to protect the coastline from destructive waves. BUT they are large and do not usually blend in with the landscape.

Rip rap: large boulders, commonly granite or concrete, placed along coastlines to absorb wave energy. BUT they are unsightly.

Revetments: often wooden frames placed at the back of a beach to absorb wave energy. BUT they need frequent maintenance.

Offshore breakwaters: a structure in the sea that is parallel to the shore which reduces the power of incoming waves. BUT they can reduce the amount of sediment reaching the beach.

Soft engineering

Some people argue that **soft engineering** is only a temporary solution and protection may be outpaced by future sea level rise.

Beach nourishment: adding sediment to a beach to increase its ability to absorb the waves and protect the land behind. BUT sediment is continually transported away, so needs constant replenishing.

Soft engineering approaches involve working with the physical processes and are a more natural form of coastal management

Cliff regrading and drainage: the angle of the cliff is reduced and water is drained out to lower the risk of mass movement. BUT this removes part of the cliff and overextraction of water can increase vulnerability of collapse.

Dune stabilisation: planting vegetation on fenced-off areas of sand dunes to hold the dune structure together and enhance the natural barrier along the coastline. BUT this disrupts the natural succession of dunes.

Key terms — Make sure you can write a definition for these key terms

dune stabilisation   environmental refugees
hard engineering   offshore breakwaters   soft engineering

3.10 Coastal management

Sustainable coastal management

Researchers predict that future global warming will put more coastal communities at risk from the threat of rising sea levels and more intense tropical storms.

- Sustainable coastal management involves protecting a significant stretch of coastline from erosion and inundation from floodwater, conserving coastal ecosystems, and ensuring the livelihoods of those who live near the coast.
- There is not a single approach that will suit all coastlines and there is conflict over which approach is best.

Holistic integrated coastal zone management (ICZM)

Many countries use sustainable, **integrated coastal zone management (ICZM)** strategies. These are **holistic** strategies because they are used to manage extended areas of the coastline within sediment cells.

ICZM:

- involves all stakeholders, taking their views and needs into account
- ensures that approaches in one area of the littoral cell does not have negative impacts elsewhere
- ensures that management is long-term, sustainable, and allows for economic development
- can change as the threats to coastal areas develop.

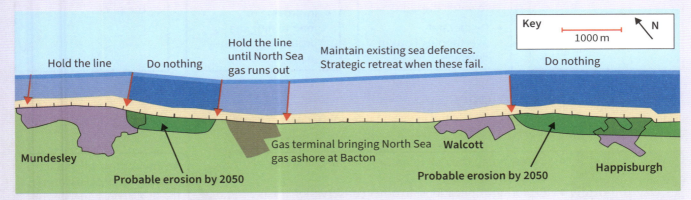

▲ **Figure 1** Holistic coastal management on the Holderness coast, Yorkshire, UK

◀ **Figure 2** Coastal erosion at Skipsea, Yorkshire on the Holderness coast

3 Coastal landscapes and change

3.10 Coastal management

Shoreline management policy

Each of the UK's 11 sediment cells is managed by a **shoreline management plan (SMP)**.

Each SMP:

- is devised by local councils and the Environment Agency, with input from other organisations
- outlines a sustainable approach to managing the threats to the coastline over 100 years
- identifies the opportunities to improve the coastal environment, the best approach to defend coastal assets and manage risks, and the consequences of putting the management in place
- works with natural processes and allows natural coastal change.

Shoreline management policy options

No active intervention	Strategic realignment	Hold the line	Advance the line
• There are no coastal defences in place.	• Moving the line of defence further towards the land. • Deliberately overtopping current defences.	• Maintaining the line of existing defences. • Improving or maintaining the protection currently in place.	• Building new coastal defences closer to the sea.

Deciding which policy to apply

Judgements to ensure that SMPs provide the most suitable policy decisions for the area consider the following aspects.

Environmental sensitivity	• Determining the risk to natural habitats and the area's biodiversity. • Exploring implications on other conservation areas such as National Parks, Sites of Specific Scientific Interest (SSSIs), Areas of Outstanding Natural Beauty (AONBs), or coastlines with heritage value.
Engineering feasibility	• Evaluating the hard and soft engineering approaches available. • Deciding if it is appropriate or possible to build manufactured structures in the area.
Land value	• Considering how the land is used. • Assessing the value of the land that is being protected in terms of economic assets.
Political context	• Applying for funding from the UK government. • Balancing the views of decision-makers in local councils with the views of local communities. These views may differ.
Social context	• Considering the social value of the coastal environments for tourism and recreation. • Recognising how local communities depend on the area for jobs, services, and wellbeing.

An Environmental Impact Assessment (EIA) assesses the likely consequences of the management activity on the environment (e.g. biodiversity, soil, water, climate, people's health).

A cost-benefit analysis (CBA) compares the economic cost of each policy (e.g. construction, maintenance) with the economic benefit (e.g. improving tourism, saving productive land).

Conflicts between different players

With any policy decisions, there will be conflict between players whose attitudes towards coastal management vary. Those who are protected or gain from the decision may be perceived as winners, and those who disagree with the decision or lose out economically or socially may be perceived as losers.

Location	Coastal management	Winners	Losers
Developed country: Happisburgh, Norfolk, UK	• There is a policy of 'No active intervention'. • The costs of constructing defences outweighs the economic benefits. • Protection at Happisburgh would accelerate coastal retreat on either side of the area.	☑ Environmental groups: the area has been designated an SSSI to ensure management does not have a negative impact on biodiversity. ☑ Tourism: areas to the south of Happisburgh will be protected by longshore drift leading to the accumulation of larger beaches.	☒ 20–35 local residents' properties will be lost to the sea. ☒ Tourism: a caravan park, beach access, churches, and a manor house will be lost by 2055. ☒ Farmers: agricultural land will be lost by 2055.
Developing country: Chattogram, Bangladesh	• Raised embankments to protect the area from rising sea levels and storm surges.	☑ Farmers: agricultural land has been protected from salt water inundation. ☑ Industry: Chattogram is an important port city and export processing zone.	☒ Residents: there have been severe traffic disruption as embankments have been raised. ☒ Some areas of the embankments are not well-maintained and are overtopped during storm surges from tropical storms.

▲ **Figure 3** *Coastal erosion in Happisburgh, Norfolk, UK*

▲ **Figure 4** *Floods in Chattogram, Bangladesh in 2023*

REVISION TIP

You need to know about the perceived winners and perceived losers of policy decisions in a developed and a developing country.

REVISION TIP

Remember to learn some specific facts about your examples and case studies. Learn specific names and places linked to facts and statistics.

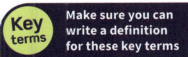

Make sure you can write a definition for these key terms

holistic   integrated coastal zone management (ICZM)
shoreline management plan (SMP)

RETRIEVAL

Learn the answers to the questions below, then cover the answers with a piece of paper and write as many as you can. Check and repeat.

	Questions		Answers
1	Name two economic losses from coastal recession.	Put paper here	Two from: properties and businesses can be destroyed / possessions of residents are lost or damaged / properties lose their value / loss of agricultural land / damage to transport and general infrastructure
2	What is an environmental refugee?		Someone who is forced from their home as a result of environmental change
3	What are hard engineering approaches?	Put paper here	Constructions such as sea walls and groynes that directly affect physical coastal processes
4	What are soft engineering approaches?		More natural forms of coastal management (e.g. beach nourishment, cliff regrading, dune stabilisation)
5	What does ICZM stand for?		Integrated coastal zone management
6	What are the four shoreline management policy decisions?	Put paper here	No active intervention, strategic realignment, hold the line, advance the line
7	What is a cost–benefit analysis?		A comparison of the economic cost of each policy with the economic benefit to determine whether a project should go ahead
8	What judgements are considered when making decisions about coastal management?	Put paper here	Engineering feasibility, environmental sensitivity, land value, political and social context
9	Who are the perceived losers in a policy of no active intervention in a developed country?		Local residents, tourism, farmers
10	Who are the perceived winners in a policy of no active intervention in a developing country?		Farmers, industry

Previous questions

Now go back and use these questions to check your knowledge of previous topics.

	Questions		Answers
1	What is the backshore?	Put paper here	The area of beach between the high tide line and the furthest extent of the beach
2	Which type of concordant coastline has lagoons created by spits that form parallel to the coast?		Haff
3	How many sediment cells are there in England and Wales?		11
4	How does a tombolo form?	Put paper here	A spit extends out to reach an island, and joins that island to the mainland
5	What are storm surges?		Short-term increases in sea level which can be caused by depressions and tropical cyclones

PRACTICE

Exam-style questions

1   Study **Figure 1**. Explain how hard-engineering approaches may alter the physical processes and systems of this coastline.   **(6)**

EXAM TIP

The Coastal landscapes and change topic is assessed in Paper 1, Section B. There will be two paragraph questions using the command word 'Explain' (6 marks), one longer paragraph question using the command word 'Explain' (8 marks), and one essay question using the command word 'Evaluate' (20 marks). You must answer all questions in the section.

▲ *Figure 1* Coastal hard engineering along the beach at Bulverhythe, Sussex, England, UK

2   Study **Figure 2**. Explain how soft engineering could protect this coastline.   **(6)**

▲ *Figure 2* Cliffs near Kessingland, Suffolk, England, UK

3   Study **Figure 3**. Explain the role of coastal deposition in the development of this landform.

(6)

▲ **Figure 3** *Cape Talumphuk in Pak Phanang, Thailand*

4   Study **Figure 4**. Explain the influence of geological structure on the development of coastal landscapes.

(6)

▲ **Figure 4** *Dorset coastline, UK*

5   Study **Figure 5**. Explain the importance of erosion processes in beach morphology.

(6)

▲ **Figure 5** *Branscombe beach, Devon, UK*

6   Explain the role of eustatic factors in causing changes in
    relative sea level. (6)

7   Study **Figure 6**. Explain the contribution of erosional processes to the
    development of this landscape. (6)

▲ **Figure 6** *Soldier's Rock, Isle of Islay, Scotland*

8   Study **Figure 7**. Explain the role of vegetation in the establishment of
    coastal landscapes. (6)

▲ **Figure 7** *An estuarine coastal landscape*

9   Explain how coastal landscapes can be classified. (8)

10 Explain how integrated coastal zone management (ICZM) strategies are
    intended to be sustainable. (8)

11 Explain the influence of geology in the development of coastal landscapes. (8)

12 Explain how bedrock lithology affects the rate of coastal recession. (8)

13 Explain the role of erosional processes in creating distinctive
    coastal landscapes. (8)

EXAM TIP

All 8 marks in these
8-mark questions are for
AO1. Make sure you show
detailed knowledge and
understanding of the
processes and concepts in
the question.

Exam-style questions

14   Explain how the coastal landscape can be viewed as a system. (8)

15   Explain the physical factors that affect coastal recession. (8)

16   Explain how coastlines can be managed using hard-engineering approaches. (8)

17   Explain why coastlines are now increasingly managed by holistic integrated coastal zone management (ICZM). (8)

> **EXAM TIP**
>
> Evaluating a view involves reviewing both sides of the argument and bringing it together to form a conclusion. Points should analyse evidence such as strengths, weaknesses, alternatives, and relevant examples.

18   Evaluate the view that coastal flood risk is mainly due to local factors. (20)

> **EXAM TIP**
>
> These 20-mark questions have 5 marks for AO1 (knowledge and understanding of the processes and concepts in the question) and 15 marks for AO2 (creating a balanced argument and well-evidenced conclusion).

19   Evaluate the view that the consequences of coastal recession and flooding are mainly economic. (20)

20   Evaluate the view that the implementation of sustainable coastal management is likely to lead to local conflicts. (20)

> **EXAM TIP**
>
> Use examples and place contexts that you have studied to exemplify your points.

4 Globalisation

4.1 Explaining globalisation

Globalisation

Globalisation is the process through which people and places across the world become more connected with each other. When different countries rely on each other, they become **interdependent**. Globalisation involves the exchange of raw materials (commodities), goods, services, capital (money), and information, as well as the movement of people (through migration and tourism).

REVISION TIP

You need to know how globalisation has affected different flows of goods, people, information, and money.

- Economic globalisation: e.g. trade flows (imports and exports of goods and services), transnational corporations (TNCs), foreign direct investment (FDI), 24/7 economy.
- Social globalisation: e.g. international migration streams, international tourism.

- Cultural globalisation: e.g. spread of ideas, information, language, e.g. English as a global language.
- Political globalisation: e.g. intergovernmental organisations (IGOs), e.g. United Nations (UN).
- Environmental globalisation: e.g. international treaties such as the Paris Agreement on climate change.

A 'shrinking world'

Advances in transport have increased globalisation. Physical distances stay the same, but travel time for goods, people, and information have rapidly decreased since the nineteenth century.

REVISION TIP

Remember to consider how developments in transport have increased trade.

Nineteenth century:
- Railway
- Steam ship
- Telegraph

→ Twentieth century:
- Jet aircraft
- Containerisation

→ Twentieth century:
- Internet
- Smartphone

Communication and globalisation

Advances in communication and technology have increased the speed of globalisation. Information can now be transferred across the globe in seconds, increasing the rate of the **time–space compression**.

Type of technology	Link to time-space compression
Mobile phones	Apps allow users fast and flexible access to a range of services from global businesses.
Internet	High volumes of data are sent around the globe instantaneously at low cost.
Social networking	Social media platforms enable the instant exchange of information.
Electronic banking	Secure financial transactions 24 hours a day, reducing the need for cash.
Fibre optics	Thin strands of glass or plastic used to transmit data, including for communication (internet), medical imaging and military operations.
Satellite technology	Receive and transmit signals around Earth. Applications include weather forecasting, real-time location (e.g. GPS), broadcasting and internet.

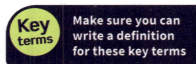 **Make sure you can write a definition for these key terms**

globalisation   interdependent   shrinking world
time–space compression

4 Globalisation

4.2 Key players in globalisation

Intergovernmental organisations (IGOs)

IGOs promote **free trade** policies to reduce barriers to trade, aim to strengthen international cooperation to improve shared prosperity, encourage foreign direct investment (**FDI**) and discourage protectionism.

World Trade Organization (WTO)

- Run by 164 member governments representing 98% of global trade
- Promotes free trade
- Negotiates trade agreements for goods, services, and intellectual property
- Oversees implementation and monitoring of agreements
- Resolves trade disputes

International Monetary Fund (IMF)

- 190 member countries
- Promotes financial cooperation
- Encourages trade and economic growth
- Discourages policies that limit prosperity
- Gives loans and financial aid to countries
- Advises on economic policy
- Provides technology and training

World Bank

- 189 member countries
- Provides financial assistance for projects for developing countries
- Focuses on reducing poverty, increasing prosperity, and supporting sustainable development

> **REVISION TIP**
>
> You need to know how free trade and FDI are promoted by the WTO, the IMF, and the World Bank.

Disadvantages of IGOs

- ☒ More economically developed countries can have more influence in decision-making, at the expense of less economically developed nations.
- ☒ Decision-making is slow and bureaucratic.

- ☒ Rules for loans can be strict, making it harder for lower-income countries to meet conditions.
- ☒ IGOs are expensive to operate.
- ☒ IGOs interfere with policies legislation of national governments, leading to sovereignty issues.

National governments

- National governments can encourage business startups to generate trade.
- Some promote free market liberalisation (removing barriers to trade between nations to encourage free trade).
- Some promote protectionism (financial support of industries within a country and the use of barriers to control imports).
- Some favour **privatisation** (when national industries such as energy, telecommunication, and rail transport owned by the state are sold to shareholders, including foreign companies).

Trade blocs

- National governments decide whether or not their country should be part of a **trade bloc**.
- Trade blocs are IGOs and have become increasingly influential to global trade.
- The WTO governs world trade by setting rules and settling disputes between countries and/or trade blocs.

REVISION TIP

Think about how governments promote trade blocs and different trade policies.

European Union (EU)

- Free trade within the bloc, and tarrifs on imported goods from non-member states.
- Four freedoms allow frictionless movement of goods, services, people, and money.
- Integration includes the use of a single currency (Euro) in most member states and EU laws, passed by the European Parliament.

Association of South East Asian Nations (ASEAN)

- Its motto, 'One vision, One Identity, One Community', promotes partnerships and growth.
- Aims to increase economic growth, social progress, and cultural development of member states.
- Free trade area, encouraging manufacturing production and financial services, including the promotion of FDI.

Advantages and disadvantages of trade blocs

Advantages (for member states)	Disadvantages (for member states and non-member states)
☑ Traded goods are cheaper because taxes and tariffs are removed or reduced.	☒ Sovereignty of member states may be limited or undermined if bloc policies do not align with national government thinking.
☑ More goods can be produced because of lower costs, leading to economies of scale.	☒ Domestic industries may decline due to cheaper competition from other member states.
☑ The movement of goods between countries is easier, as there are fewer customs restrictions.	☒ Protectionism makes it more difficult for less economically developed countries outside the bloc to access lucrative markets within the bloc.
☑ It is easier for transnational corporations (TNCs) to operate and expand within the trade bloc.	☒ Gaining agreement from all member states for new policies can be slow.
☑ Member states are part of a larger market for the exchange of goods and services.	☒ There is a perceived lack of accountability, as decision-makers have not been directly voted for by national populations.
☑ Industries within the trade bloc are protected from competition from non-member states.	
☑ Deeper policy integration can lead to more economic power and increased political security.	

4 Globalisation

4.2 Key players in globalisation

Spread of globalisation to new global regions

Since the 1960s, governments of emerging economies have encouraged investment from foreign countries and companies (FDI), to aid their economic growth and their global competitiveness, by setting up special economic zones (**SEZs**).

- SEZs are designated areas which have different rules from other parts of the country.

- International companies setting up in SEZs can benefit from no or low taxes. Tariffs on importing and exporting goods can be removed.

- SEZs are often near the coast to take advantage of trade through deep water ports, and are linked to transport and other infrastructure networks.

- SEZs act as a catalyst for economic growth within a region.

- Governments can also offer financial incentives, e.g. **government subsidies**, to encourage companies to set up, reducing the start-up costs for new businesses.

> **REVISION TIP** ☑
>
> You need to know how regions have become globalised by SEZs, government subsidies, and FDI.

Attitudes to FDI

- ASEAN actively encourages FDI (in 2022, UK firm Dyson invested $1.1bn in Singapore).

- China has increasingly received FDI, following the opening up of its economy in 1978 through the Open Door Policy.

- Due to its recent economic growth, China itself now provides FDI to over 150 countries through its Belt and Road Initiative (BRI), which aims to increase global trade by developing major infrastructure projects.

▲ **Figure 1** *Belt and Road Initiative bridge under construction in Yunnan, China*

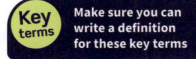

Key terms Make sure you can write a definition for these key terms

FDI   free trade   government subsidies
IMF   privatisation   SEZ   trade bloc   World Bank   WTO

4.3 Levels of globalisation

Measuring globalisation

- Some countries are more globally interdependent than other countries, and some places have benefitted more from globalisation than others.
- The level of global interconnectedness can vary both between countries and within countries.
- There are many economic, social, political, technological, and environmental factors that contribute to the level of globalisation. These factors may be external or internal to a country or place.

AT Kearney Global Cities index

The **AT Kearney index** combines and then ranks data linked to:

- business activity (e.g. top global services companies)
- human capital (e.g. top universities)
- information exchange (e.g. broadband subscribers)
- cultural experience (e.g. sporting events)
- political engagement (e.g. embassies and consulates).

Between 2017 and 2022, Beijing rose from nineth to fifth in the rankings. Between 2021 and 2023 the rank order of the top four cities remained the same:

1. New York
2. London
3. Paris
4. Tokyo

KOF index

- The **KOF index** measures globalisation by a combination of economic, social, and political factors.
- The range of data included is complex, and includes the amount of trade of goods and services, the level of international debt, gender parity, and internet access.
- Countries with a high ranking are more globalised than those with a low ranking.

Most globalised countries in 2023 based on KOF index		Least globalised countries in 2023 based on KOF index	
Rank	Country	Rank	Country
1	Switzerland	192	Afghanistan
2	Netherlands	193	West Bank and Gaza
3	Belgium	194	Central African Republic
4	Sweden	195	Eritrea
5	Germany	196	Somalia

Using indicators to measure globalisation

Advantages	Disadvantages
☑ Easy to compare countries	☒ Do not show differences between regions within a country
☑ Easy to see trends (changes over time)	☒ Data may not be reliable, e.g. problems with data collection
☑ Easy to recognise patterns (differences between places)	☒ Data may not be valid, e.g. out of date
☑ Compares a range of factors, giving a more rounded view	☒ Some variables use qualitative data and so can be more subjective than variables using objective data

4 Globalisation

4.3 Levels of globalisation

Transnational corporations (TNCs)

TNCs aim to increase profits and market share by developing different parts of their business in cost-effective locations and in developing markets.

Company headquarters
Quaternary employment (e.g. R&D, human resources, operations, finance, legal, sales and marketing, ICT)
• Highly-skilled staff
• High wages
HIC (Global North)

Global production networks (GPNs)
Secondary employment (e.g. manufacturing, assembly)
• **Outsourcing** and/or **offshoring**
• Lower wage rates in emerging economies
MIC/LIC (Global South)

Markets
Tertiary employment (e.g. retailing and e-tailing)
• Access to growing markets, UMICs, and LMICs
HIC/MIC and some LICs

Raw materials suppliers
Primary employment (e.g. mineral extraction, crop production)
• Availability of cheap raw materials
HIC/MIC/LIC

TNCs and globalisation

Glocalisation

Glocalisation is when a company changes some of its products to suit the local market so that sales increase. TNCs use glocalisation to access new markets.

Development of new markets

- BRICS countries (Brazil, Russia, India, China, and South Africa) and MINT countries (Mexico, Indonesia, Nigeria, and Türkiye) provide large and/or growing markets, in terms of population and personal wealth.
- Challenges in these markets include corruption, conflicts, economic crises, and climate-related issues.

Economic liberalisation

TNCs have benefitted from economic liberalisation, which promotes free trade between countries and the reduction of government intervention in the market. This stimulates an increase in wealth, which should 'trickle down' to other parts of the economy and across the population.

Outsourcing

Outsourcing is when a TNC subcontracts the manufacture of its products or the delivery of its services to another business (third party).

Advantages	Disadvantages
☑ Fewer set-up costs for TNC	☒ TNC may move to a cheaper location, leading to local job losses
☑ Can move production more flexibly, if needed	☒ TNC's brand reputation may be at risk if subcontractor does not follow correct production methods or local labour/environmental laws
☑ Can develop local links with connected businesses	
☑ Access to local markets	
☑ Takes advantage of lower labour costs	☒ Natural disasters can affect production, labour supply, transportation, and markets
☑ Can stimulate the multiplier effect	☒ Supply chains affected by pandemics or political instability
☑ Can avoid tariffs if producing and then selling within a trading bloc	

Offshoring

Offshoring is when a TNC does work overseas, either itself or using another company.

Advantages	Disadvantages
☑ Keeps control of its operations within the company whilst lowering costs	☒ Language and communication barriers (e.g. local accents, different time zones)
☑ Low labour costs, including for higher skilled labour (e.g. wages)	☒ Cultural differences (e.g. social customs)
☑ Other cost savings (e.g. electricity)	☒ Quality control is more difficult
☑ Tax benefits (e.g. tax incentives and tax holidays)	☒ Loss of jobs in home countries
☑ Profit increase (e.g. revenue returned to home country)	
☑ 24/7 operations (e.g. takes advantage of time zones)	

'Detached' locations

These locations have limited access to global markets and networks. This leads to:

- no or low levels of foreign investment
- low volumes of, and income from, international trade flows
- limited sharing of ideas and information with the 'outside world'.

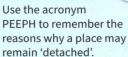

REVISION TIP

Use the acronym PEEPH to remember the reasons why a place may remain 'detached'.

Reasons why locations remain 'detached'

Political
- Government and its policies
- Barriers to migration
- Lack of influence within IGOs
- External and/or internal conflicts

Economic
- Lack of investment in infrastructure and technology
- Poor management of natural resources
- Lack of access to global trading system
- Debt and debt-servicing burden

Environmental
- Risk of impacts of climate change
- Threat of pests and disease
- Poor management of natural hazards

Physical
- Landlocked
- Global periphery, remote
- Extreme climates
- Natural hazard risks
- Uninhabitable terrain

Historical
- Conflicts and disputes may lead to countries isolating themselves to protect their own sovereignty
- Countries may have had their resources, including workers, exploited by former imperial powers
- Countries may have allied themselves with other countries that had similar political ideologies but have since changed

North Korea: a 'detached' country

The main reasons why North Korea is a **'detached' country** are political:

- North Korea has a one-party communist system of government and is run through a command economy (government, rather than the market, dictates what and how much is produced).

- There are low levels of official trade between North Korea and other countries. Foreign tourism is highly restricted and emigration is forbidden.
- Internet connections are only permitted with special authorisation.

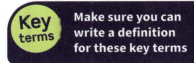

Key terms Make sure you can write a definition for these key terms

AT Kearney index   detached country   KOF index
offshoring   outsourcing   TNC

Learn the answers to the questions below, then cover the answers column with a piece of paper and write as many as you can. Check and repeat.

Questions

Answers

	Questions	Answers
1	What is globalisation?	The process of increased interconnectedness and interdependency between places and people
2	What is foreign direct investment (FDI)?	Money invested by a foreign country or company (TNC) into a country
3	Name one type of factor that can be used to measure the level of globalisation.	One from: economic / social / political / environmental / technological
4	Name one measure of globalisation.	One from: AT Kearney Global cities Index / KOF index
5	Give one advantage of using indicators to measure globalisation.	One from: easy to compare countries / easy to see trends (changes over time) / easy to recognise patterns (differences between places) / compare a range of factors, giving a more rounded view
6	Give one disadvantage of using indicators to measure globalisation.	One from: do not show differences between regions within a country / data may not be reliable (e.g. problems with data collection) / data may not be valid, (e.g. out of date) / some variables use qualitative data and so can be more subjective than variables using objective data
7	Name the four key components of a TNC's global operations.	Company headquarters, raw materials suppliers, global production networks, market
8	What is outsourcing?	When a TNC subcontracts part of its business to another company
9	What is offshoring?	When a TNC moves part of its operation to another country to decrease its costs
10	What is a 'detached' location?	A place which does not have access to global markets and networks
11	Give one economic reason why a place may be 'detached'.	One from: lack of investment in infrastructure and technology / poor management of natural resources / lack of access to global trading system / debt and debt-servicing burden
12	Give one environmental reason why a place may be 'detached'.	One from: risk of impacts of climate change / threat of pests and disease / poor management of natural hazards
13	Give one physical reason why a place may be 'detached'.	One from: landlocked / global periphery, remote / extreme climates / natural hazard risks / uninhabitable terrain
14	Give one historical reason why a place may be 'detached'.	One from: isolation following conflicts and disputes / previous imperial resource exploitation / previous alliances with other countries with similar political ideologies that have changed

Put paper here

⚙ KNOWLEDGE

4 Globalisation

4.4 Impacts of globalisation

The global shift

- In the late twentieth century, the **global economic centre of gravity** started to move towards Asia, as manufacturing moved from Europe and the USA to Asia – the **global shift**.
- Asia's share of global GDP has risen since the 1980s and is predicted to increase further into the twenty-first century.
- FDI is also an indicator of economic change – Asia now has the highest levels of FDI in the world.

After Second World War >	Since 1960s >	Since 2000s >
Asian countries develop their manufacturing industries to grow their economies.	Japan, and then the so-called 'Asian Tigers' (Singapore, Hong Kong (now part of China), South Korea, and Taiwan), achieve high levels of economic growth by rapidly developing their industrial sector, producing exports for the global market.	Asian companies are highly competitive in the global hi-tech sector.

China

- A large population means high numbers of workers are available. Many have moved from rural to urban areas to work in manufacturing centres on the coast, e.g. Shenzhen.
- Initially, conditions in factories were poor. But, during the 2000s, conditions improved slightly, including better wages, leading to an improvement in living standards for many.
- Initially, products were lower-value manufactured goods. But, since the 2010s, quantities of higher-value products (e.g. iPhones) have increased.

- Chinese technology companies have developed products for the global market (e.g. Huawei smartphones) and bought into social media businesses (e.g. TikTok).
- As wages in China increase, some companies are seeking manufacturing bases in other economies in Southern Asia with lower wages.
- Since 1990, China's GDP growth has averaged around 9% per year – much higher than the rest of the world.

Outsourcing of services

- TNCs operate and grow in international markets, so need different services to support their activities 24 hours a day, 7 days a week.
- To reduce labour costs, TNCs set up operations in emerging economies where the labour force's skills match the company's needs, enabled by the development of telecommunications and internet-based technology.

India

- Since the 1990s, businesses have set up in technology parks offering incentives such as low-cost rates.
- There is access to a large population with a well-educated workforce.
- Many employees are willing to work long hours, including night shifts, which helps employers working in different time zones.

- Some employees can earn good wages compared to other workers in India, increasing their standard of living. Their spending contributes to the multiplier effect.
- Internal migrant workers contribute to the economy of the region they moved away from by sending money home.
- The percentage of people living in extreme poverty has reduced from 50% in 1990 to 10% in 2021.

4 Globalisation

4.4 Impacts of globalisation

Growing inequalities

The global shift has also created growing inequalities. The rise of TNCs in India has tended to polarise growth in urban areas with some rural regions being disadvantaged. Increasing rates of poverty and low incomes in agriculture led to farmers in rural India protesting in the early 2020s. Those on lower incomes in rural areas may find it difficult to access health and education, creating a cycle of poverty and increasing inequalities between regions. Access to key technology (e.g. the internet) may cause a 'digital divide' between urban and rural regions.

Environmental impacts

Air and water pollution

High emissions of carbon dioxide and other gases due to manufacturing processes lead to a reduction of air quality. In 2021, a number of schools in Beijing temporarily closed due to high air pollution levels.

Land degradation

- Growing manufacturing hubs, and urban development to house the increased workforce, lead to deforestation and the loss of productive agricultural land.
- Land quality is reduced through the release of toxic chemicals and the dumping of waste products in the manufacturing process.
- Land on the urban periphery is used by homeless migrant employees for informal housing settlements.

Over-exploitation of resources

The increased demand for resources such as energy, water, and minerals, leads countries to import resources.

Loss of biodiversity

Loss of habitats reduce species numbers and disrupt the balance of fragile ecosystems. Complex food chains can be disrupted, leading to the collapse of species' populations.

Deindustrialisation in developed countries

Industrial land no longer in use becomes derelict. Land is contaminated and expensive to clean up. Some sites (e.g. dock areas) present physical challenges for redevelopment to other uses.

Unemployment rates (and so poverty) increase. Lack of transferable skills makes it difficult for redundant workers to find jobs in newer industries. Economically active people may move to other places. Younger workers may seek opportunities elsewhere.

Social and environmental problems as a result of economic restructuring

Urban areas become depopulated and house prices decline as few people want to move into these areas.

There is a spiral of decline leading to lower or no wages, poverty, crime, and antisocial behaviour. Governments have to offer incentives for businesses to set up in these areas.

These regions have to encourage investment in new types of economic activities to provide income and opportunities for employment.

 Key terms Make sure you can write a definition for these key terms

economic restructuring   global economic centre of gravity   global shift

4.5 Migration

The growth of megacities

- In 1970, there were three cities with populations over ten million (**megacities**): Tokyo, New York, and Osaka.
- Migration, including **rural–urban migration**, and **natural increase** mean that cities are rapidly growing into megacities.
- By 2030, estimates suggest there could be over 40 megacities. The majority of these will be in South and East Asia.

REVISION TIP

Practise listing the causes of growth of megacities from memory.

Push factors (why people leave rural areas)	Pull factors (why people move to urban areas)
<ul><li>Increased mechanisation means fewer agricultural workers are needed.</li><li>The reduction in productive land owned by local people, due to land reforms and purchase of land by TNCs.</li><li>Climate change is affecting rainfall quantity and frequency, leading to land degradation.</li><li>Fewer employment opportunities for increasingly educated and skilled young workers.</li><li>The lack of services (e.g. healthcare and schools) in remote rural areas.</li><li>The lack of infrastructure, leading to rural areas being 'detached'.</li><li>The threat of conflict or risk of natural disasters.</li></ul>	<ul><li>A perception of increased life chances and standard of living.</li><li>Higher paid employment opportunities, often due to economic investment (e.g. FDI).</li><li>Better access to and greater variety of health and education services.</li><li>Better IT connectivity.</li><li>More social opportunities, especially for younger people.</li><li>More reliable food supply.</li><li>The opportunity to be near family and friends who have previously migrated (chain migration).</li></ul>

Challenges of urban growth

Social

- Lack of sufficient employment opportunities
- Rise of informal settlements on edge of cities
- Inadequate housing standards
- Lack of affordable housing
- Overcrowding
- Lack of (legal) access to regular supply of water/electricity, sanitation, internet
- Higher crime rates; possible tensions between some migrants from different communities

Environmental

- Water pollution, including from inadequate sanitation (sewage), industrial waste, and non-reusable plastics
- Air pollution, including from increased traffic and industrial processes
- Noise pollution; 24/7 city means more activity, including traffic throughout the day
- Land degradation, including from toxic waste
- Encroachment on wildlife habitats, leading to loss of biodiversity and conflict

4 Globalisation

4.5 Migration

Global hub cities

A **global hub** is a city that is highly connected globally, through **international migration**, business and cultural links, and transport infrastructure. These connections deepen interdependence between countries.

The growth of global hubs has been driven by:

REVISION TIP

List the effects of international migration on hub cities and regions.

- **Elite international migrants**: highly skilled, influential, and affluent people, often seen as 'global citizens', with limited obstacles to migration. They can take advantage of differences in tax and residency rules between countries. For example, high-wage economic migration to London and Singapore.

- **Low-waged international economic migrants**: these workers work in support and ancillary services in global hubs (e.g. in domestic help, construction, and security). Workers can have legal or illegal status. For example, people from India moving to the UAE or from the Philippines to Saudi Arabia.

Benefits and costs of migration

Migration impacts both the source country or region (the place the migrant leaves) and the host country or region (the migrant's destination).

Benefits for the source country

- Receives income from money sent home.
- Migrants are likely to have new/developed skills if they return.
- Less expenditure on services, e.g. health.
- Increased jobs available for those who remain.
- Stronger ties between nations due to the diaspora (links between migrants and their country of origin).

Benefits for the host country

- Migrants help fill labour shortages in both the less-skilled (e.g. agricultural, hospitality) and higher-skilled (e.g. healthcare) sectors.
- Migrants spend money in the local economy.
- Increased economically active population reduces proportion of dependent population (e.g. ageing population).
- More diverse and cosmopolitan society.
- Dynamic migrants develop entrepreneurial skills.

Costs for the source country

- 'Brain drain', as skilled workers move elsewhere to earn money, including health, education, and business professionals.
- Decline in birth rate and overall population.
- Increase in ageing population with potential need for healthcare.
- Rural depopulation increases.
- Closure of businesses and services aimed at the younger section of the economically active population.
- Family units divided if one parent or carer works in another country.

Costs for the host country

- Increase in social tensions as citizens and migrants both need access to affordable housing, school places, and healthcare.
- Increase in political tensions, including parties promoting the reduction of migration, e.g. Brexit.
- Difficulties of communication and access to services caused by language barriers.
- Marginalisation of ethnic communities.

REVISION TIP

Compile a table to show the costs and benefits of migration for both host and source locations.

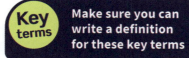

Key terms

Make sure you can write a definition for these key terms

elite migration   global hub city   low-wage migration   megacity
natural increase   pull factor   push factor   rural-urban migration

4.6 Global culture

Cultural diffusion

- Some believe that globalisation has led to **'westernisation'** of cultures around the world. This is **cultural diffusion**.
- Cultural imperialism occurs when one country's culture is promoted ahead of another. For example, English is an official language in some countries that were colonised by Britain.
- Cultural change can also occur through 'soft power' where business interests, education, media, and entertainment alter people's way of life.

Role and impact of different players in cultural diffusion

TNCs

- The largest and most influential TNCs tend to be from the USA and Europe.
- Their influence can be through desirable products, business organisation, and working practices.

Media corporations

- Information can now be transmitted instantly. Social media contributes to the diffusion of culture and ideas.
- Media corporations access a global market, spreading 'western' ideas.
- However, increased availability of international films and TV can expose people to different cultures.

Tourism

- Tourists can bring their own behaviours and languages to places they visit.
- Clashes can occur, e.g. intolerance towards LGBTQIA+ people; different attitudes about alcohol use and styles of clothing.

Migration

- Migration can lead to cultural changes, as migrants adapt to their new societies.
- Diverse communities may live together, or there can be cultural segregation.
- Migration can lead to 'third-culture kids', where children are raised in a culture different from their parents or carers.
- Migration can influence the built environment, e.g. new religious buildings for a growing immigrant community.

Changing diets in Asia

Since 1961, the daily calorie intake in Asia has increased due to global trade.

Impact on health and environment

- Increasing consumption of red meat, dairy products, and processed and fast food has led to obesity, cardiovascular diseases, and diabetes.
- Changes in agricultural land use, from cereal production to animal production, have increased carbon dioxide and methane emissions.
- Asia has the largest share of the world's population meaning that global food production must increase.

Global culture and opportunities for disadvantaged groups

- IGOs and national governments have been involved in improving life for disabled people.
- Paralympic Games: in 1960 (Rome), 400 athletes from 23 countries took part; in 2016 (Rio de Janeiro), 4342 athletes representing 159 countries took part.

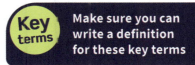

Key terms Make sure you can write a definition for these key terms

anti-globalisation   cultural diffusion   cultural erosion
pro-globalisation   westernisation

4 Globalisation

4.6 Global culture

Cultural erosion

Loss of language: English has become the dominant global language.

Traditional foods: eaten less; recipes risk being lost.

Music: pop-music culture is global; traditional music may decline.

Cultural erosion

Social relations: migration means family units are more disparate.

Clothes: global fashion and sportswear brands instead of traditional clothes.

In some places cultural erosion is the result of deliberate action

Loss of indigenous lifestyles in Papua New Guinea

- The discovery of valuable minerals has led some indigenous communities to change from subsistence forest-based agriculture to resource extraction.
- Growing demand for timber has caused deforestation, reducing biodiversity and the area available to find traditional food sources.
- Toxic waste from large-scale mining projects and increased surface runoff has impacted river and ocean water quality.
- Community based organisations aim to protect primary rainforest, reduce the rate of deforestation, and promote sustainable traditional livelihoods.

> **REVISION TIP**
>
> Think about examples with locations of how globalisation erodes culture.

Changes to the built and natural environment

Built environment	Natural environment
• Skyscrapers are common in the financial districts of global hubs where land prices are high. • New developments, including in developing and emerging economies, are often modern in style to attract TNCs.	• Altered by economic change and urban development. • Mass tourism can impact land- and marine-based ecosystems. • Construction of hotels and provision of resources for tourist needs (e.g. water) have a negative effect on local ecosystems. • Tourist plane travel increases carbon dioxide emissions.

UNESCO World Heritage Sites have been designated to protect culturally important landscapes and heritage sites, e.g. Kinabalu Park in Borneo (Malaysia).

Support for and opposition to globalisation

Anti-globalisation groups may believe that:

- globalisation leads to employment loss and economic inequality
- TNCs and IGOs have polarised development
- the spread of 'western' ideas and practices poses a threat to different cultures.

Pro-globalisation groups believe that:

- globalisation has helped to improve connectivity between nations
- connectivity has led to an increase in living standards and a decrease in poverty.

> **REVISION TIP**
>
> Name examples of groups that have different views on globalisation. Write a sentence for each to say why they are for or against globalisation.

⇄ RETRIEVAL

Learn the answers to the questions below, then cover the answers column with a piece of paper and write as many as you can. Check and repeat.

Questions | Answers

	Questions		Answers
1	What is the global shift?		Movement of manufacturing from the USA and Europe to Asia
2	What is land degradation?	Put paper here	Reduction of quality of land over time
3	What is the term for a reduction of plant and animal life in a given area?		Biodiversity loss
4	What is economic restructuring?		Changes in types of economic activity over time in a given area
5	What is a megacity?	Put paper here	A city with over ten million people
6	Give the two main reasons for the growth of megacities.		Migration (including rural–urban migration), natural increase
7	What is a push factor?		Reason for leaving a particular location
8	What is a pull factor?	Put paper here	Reason for moving to a particular location
9	What is an elite international migrant?		A wealthy, influential migrant
10	What is cultural imperialism?		The imposing of one country's culture on another

Previous questions

Now go back and use these questions to check your knowledge of previous topics.

Questions | Answers

	Questions		Answers
1	Name one type of factor that can be used to measure the level of globalisation.		One from: economic / social / political / environmental /technological
2	What is a 'detached' location?	Put paper here	A place which does not have access to global markets and networks
3	What is outsourcing?		When a TNC subcontracts part of its business to another company
4	What is offshoring?	Put paper here	When a TNC moves part of its operation to another country to decrease its costs
5	Give one physical reason why a place is 'detached'.		One from: landlocked / remote / extreme climates / natural hazard risks / uninhabitable

4 Globalisation

4.7 The development gap

Economic measures of development

Single indices, such as GDP, GDP per capita at PPP, and employment in services (percentage of total employment).

Advantages	Disadvantages
☑ Straightforward to understand, including for a non-specialist audience	☒ Oversimplified view of the economy
☑ Released regularly by governments and organisations	☒ Economic indicators are often interconnected, so meaningful trends may not be identified easily
☑ Used to focus on a particular aspect of the economy, helping planning and response	☒ Money values are quickly affected by external shocks, so data can be inaccurate
☑ Allows for comparison with historic data	☒ Some countries may not have the resources to collect accurate data

Composite indices, such as the OECD's Composite Leading Indicators (CLI).

Advantages	Disadvantages
☑ More comprehensive, holistic view	☒ More complicated to understand, especially for a non-specialist audience
☑ More complex data can increase validity	☒ Data for each component may not be available for the same timeframe or level of accuracy for different countries, making comparison more difficult
☑ Reduces bias towards one particular indicator when making a judgement	
☑ Gives decision-makers a better understanding of economic conditions	☒ Weightings given for each component are subjective

Social measures of development

Human development index (HDI)

- The HDI shows how far people are benefitting from economic growth.
- It uses numbers between 0 (lowest HDI) and 1 (highest HDI) to give a summary measure of average achievement in key dimensions of human development.
- HDI dimensions: long life (life expectancy in years); living standard (GBP per capita); education (literacy rate and average years of schooling).

Gender inequality index (GII))

- The GII is measured using a scale from 0 (lowest inequality) to 1 (highest inequality), obtained from an average of the score for each dimension.
- GII indicators: reproductive health (maternal mortality ratio and adolescent birth rate); empowerment (females with secondary or higher education, share of females in parliament); labour market (female participation in labour force).

Environmental quality (air pollution indices)

- Air quality deteriorates as economic development increases. Industrial activity and transportation emit gases and particulates.
- Air pollution can lead to respiratory problems such as asthma.

- Most HICs try to control vehicle emissions and regulate industrial pollution (or outsource the industry and the air pollution problem).
- Air pollution indices show the level of real-time air quality and predict future air quality.
- Different countries have different types of air quality indices, making direct comparisons difficult.

Widening income inequality

- Globalisation has led to rises in incomes and quality of life, but the rate of change is not equal across world regions.
- Wealth can vary between and within countries.
- Absolute poverty has declined, but relative poverty has increased, including in some HICs such as the UK.

- Economic inequality within a country can lead to social issues and political unrest.
- Globalisation has also affected the physical environment: urban sprawl has increased; land degradation caused by intensification of agriculture and increases in monocultural plantations is widespread.

Measuring inequality: the Gini coefficient

- Measures the level of income inequality within a country, focusing on the distribution of household income
- Values range between 0 (no inequality) and 100 (highest inequality).
- It is calculated from the Lorenz curve of household income distribution.

REVISION TIP

Practise explaining to a classmate what the Gini coefficient measures.

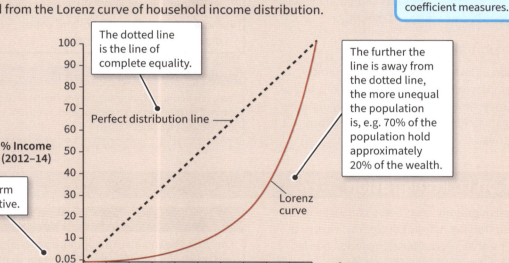

The dotted line is the line of complete equality.

Perfect distribution line

The further the line is away from the dotted line, the more unequal the population is, e.g. 70% of the population hold approximately 20% of the wealth.

% Income (2012–14)

Data is in % form and is cumulative.

Lorenz curve

Decile of population

◀ **Figure 1** The Lorenz curve, used to plot and calculate the Gini coefficient

Trends in economic development and environmental management

- There has been an increase in the profile and awareness of the need to reduce the environmental impact of industrial development.
- IGOs and HICs have been at the forefront of environmental regulation and legislation. For example, the UK has rules such as the 'polluter pays principle' and carbon offsetting schemes.
- Critics argue that LICs are being financially penalised for industrialising later than HICs, and having to pay either directly or indirectly for meeting stricter global environmental legislation.
- Successful environmental management has economic and social benefits, including the effect on the tourism sector, an important source of GDP and employment in many LICs.

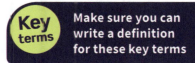

Key terms Make sure you can write a definition for these key terms

Gender Inequality Index (GII)   Gini coefficient
Human Development Index (HDI)   Lorenz curve

4 Globalisation

4.8 Tensions over globalisation

Growing diversity

Open borders

Open borders are those across which there are fewer restrictions on the movement of people, goods, and services, money and information.

In 2004, ten countries joined the EU in the largest expansion in its history. Over 800,000 Polish people moved to the UK. Polish migrants settled in higher concentrations in some areas of the UK.

Deregulation and FDI

Deregulation allows FDI to come into a country, providing employment and developing skills and expertise. Governments may also relax regulations to allow organisations to employ workers from other countries.

Boston, Lincolnshire

- Overseas population increased by 470% between 2001 and 2011.
- Eastern European migrants have helped resolve the skills shortage in the agricultural industry.
- Workers spend money in local shops and businesses, and a Catholic church and Polish shops have opened.
- Increased need for school and college places and demand for health services.

REVISION TIP

Make an audio recording explaining how globalisation has caused tension between groups and listen to it when you are travelling or waiting.

Opposition to immigration

International migration has created culturally mixed societies and thriving migrant diasporas. However, some perceive that international migration has led to:

- a loss of national identity and cultural norms and values
- pressure on school places and increased demand for health services
- a need for more affordable housing
- more competition for jobs.

Rise of extremism

There has been a rise in popularity of nationalist parties and movements in Europe. They want to prioritise national interests over global ones and favour restrictions on immigration. They are often supported by groups that feel marginalised by international migration, e.g. those on low incomes.

Trans-boundary water conflicts in South-East Asia

- The Tibetan Plateau and Hindu Kush-Himalayan region supplies nearly 2 billion people (25% of the global population) with water. Projections suggest that water demand in the region will increase 11% by 2050.
- Climate change has caused the region to warm at rates higher than the global average and precipitation patterns have changed.
- Short term, glacial meltwater will increase the total water volume, increasing the likelihood of flooding.
- Long term, a lack of replenishment of glaciers will lead to a decrease in overall water supply, particularly in the south of the region.

India

- Has raised concerns about the construction of dams in the Tibet region, part of the Brahmaputra drainage basin.
- Is building a $1.2bn Pakal Dul HEP facility on the Marusudar River, which could increase India's water security, but negatively impact Pakistan's.

China

- Is seen as an 'upstream superpower' yet has not signed international water resource agreements with its neighbours.
- Has developed green energy and flood control through building dams, but this impacts the amount of water reaching other countries.

Controlling globalisation

Censorship

The World Press Freedom Index (2023) ranks North Korea as having the least press freedom in the world, after China. The Chinese authorities perceive criticism of their policies as a threat and try to prevent this by monitoring and controlling the media.

Limiting migration

Since Brexit, the UK government has brought in schemes to control migration to the UK. New migrants need to have permission to stay, e.g. a work, family, or student visa. A points-based system prioritises those with certain skills and a higher salary. EU citizens must now apply for the EU Settlement Scheme to stay in the UK.

Trade protectionism

Governments can impose tariffs or quotas on imported goods, or give subsidies to financially protect their own industries from fluctuations in value of goods on the global market. However, the WTO promotes free trade and members must abide by the WTO rules.

> **REVISION TIP**
>
> Compile a table to help you remember how different players have tried to control the spread of globalisation.

Retaining cultural identity within countries

First Nations Communities in Canada

Energy TNCs have extracted oil and gas from areas that First Nations communities (indigenous peoples) live in.

Development has:

- brought employment and increased income
- increased water pollution and habitat degradation
- changed traditional lifestyles.

This has caused conflict with First Nations communities.

Retaining control of culture and physical resources

First Nations communities have become more involved with the decision-making around development.

- Property rights have been strengthened and First Nations communities are consulted before economic development is allowed on their land.
- First Nations communities have become small-scale investors, e.g. Fort McKay and Mikisew Cree First Nations invested in Suncor East Tank Farm in Alberta.

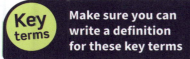

Key terms Make sure you can write a definition for these key terms

censorship   cultural identity   deregulation   extremism
open borders   trade protectionism

4 Globalisation

4.9 Ethics and environmental concerns

Local sourcing as a response to globalisation

Transition towns, part of the global Transition movement, aim to achieve low-carbon, socially just, resilient societies, with a caring culture and active participation amongst residents. Projects include promoting renewable energy, creating local currencies, maintaining local food systems, and creating community and green spaces.

Transition town: Totnes, Devon, UK

A community-led charity that promotes:

- community gardens (growing fruit and vegetables on common land)
- shared office space where new and existing enterprises can develop their businesses, knowledge, and skills, and collaborate on projects
- bringing neighbours together to make streets more connected
- educating the community to be more sustainable.

Advantages of local sourcing	Disadvantages of local sourcing
☑ Directly supports local farmers and businesses, circulating money in the local economy and sustaining local employment	☒ No economies of scale, so products are more expensive
☑ Products do not travel far, so carbon footprint is reduced	☒ Smaller volume of goods produced, so supply might not meet demand
☑ Food products are fresher	☒ Food products can be seasonal, so not available all year
☑ Local traditions in craft-making are sustained	☒ Less variety of products may be available
☑ Farmland in local areas is preserved, maintaining greenspace	☒ May affect producers in LICs who are less able to access markets
☑ Food security is improved as the local population maintains its food supply if global supply chains are disrupted	

Ethical consumption schemes and fair trade

Ethical consumption

Ethical consumption is the buying and use of goods and services that minimises the damage to society and the environment caused by their production. It allows consumers to take more responsibility and become more engaged with sustainable development.

Fair trade

- Organisations such as Fairtrade International guarantee a minimum price for farmers even if the price of the product on the market falls, as well as an additional fair trade premium, giving farmers a sustainable livelihood.
- Fair trade also protects workers' rights and upholds environmental standards.
- Products are given a fair trade mark so the consumer knows that they have been produced responsibly and the producers have received a fair price for their product.

> **REVISION TIP** ☑
>
> Draw a flow diagram to show how fair trade works, from farmer to supermarket shelf.

Benefits of recycling and the circular economy

Conservation of resources through **recycling** materials rather than extracting and processing new raw materials.

Development of new greener technologies for recycling waste reduces the **ecological footprint** and creates employment in the sector.

Reduces landfill waste, reduces the need for new sites, and extends the lifespan of existing sites.

Recycling requires less energy than producing goods from raw materials.

Resources are kept in use as long as possible in the production and consumption cycle.

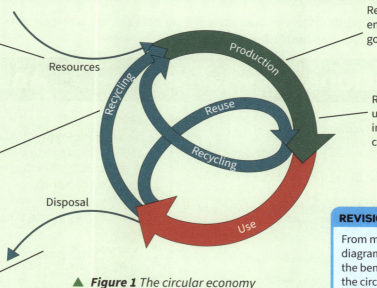

▲ **Figure 1** *The circular economy*

> **REVISION TIP**
>
> From memory, draw a diagram that explains the benefits of recycling and the circular economy.

Local initiatives

Keep Britain Tidy

Strategies of this UK NGO include: community litter picks, campaigns to change behaviour, promotion of reuse and recycle, the Green Flag award for parks and green spaces, and educating children on sustainable actions.

Local Agenda 21 (LA21)

This sustainable development initiative, promoting environmental programmes and community involvement at a local level, originated from the UN Earth Summit in Rio 1992. LA21 has been incorporated into a range of UK local government strategies.

Recycling success varies

Recycling success varies between products and different locations due to a range of factors:

- Access to frequent recycling collection and recycling facilities.
- Clear guidelines and standards on what can and cannot be recycled.
- Public awareness and education.
- Effectiveness of recycling facilities to control contaminated materials.
- Cost of setting up and maintaining recycling schemes.
- Technological development increasing the volume and type of products that can be recycled.
- Encouraging personal, community, and corporate engagement in recycling.

 Key terms Make sure you can write a definition for these key terms

ecological footprint   fair trade   local sourcing
recycling   transition town

Learn the answers to the questions below, then cover the answers column with a piece of paper and write as many as you can. Check and repeat.

Questions | Answers

	Questions		Answers
1	Name a composite economic measure of development.	Put paper here	The OECD's Composite Leading Indicators (CLI)
2	What is the Human Development Index (HDI)?		A composite index measuring health, education, and standard of living
3	What are the three dimensions that contribute to the HDI?		Long life, living standard, education
4	What is the Gender Inequality Index (GII)?	Put paper here	A composite index measuring reproductive health, empowerment, and labour
5	Why is it difficult to directly compare air quality indices between countries?		Different countries calculate air quality indices in different ways
6	State one way in which globalisation can affect the physical environment.	Put paper here	One from: increased urban sprawl / land degradation caused by intensification of agriculture and increases in monocultural plantations
7	What does the Gini coefficient measure?		Income inequality within a country
8	What are the three key ways to control globalisation?		Censorship, limiting migration, trade protectionism
9	What is local sourcing?	Put paper here	Goods and services are made within or obtained from the local community
10	What is ethical consumption?		The buying and use of goods and services that minimise the damage to society and the environment caused by their production

Previous questions

Now go back and use these questions to check your knowledge of previous topics.

Questions | Answers

	Questions		Answers
1	What is outsourcing?	Put paper here	When a TNC subcontracts part of its business to another company
2	What is foreign direct investment (FDI)?		Money invested by a foreign country or company (TNC) into a country to build a new part of their business
3	What is a megacity?		A city with over ten million people
4	What is a push factor?	Put paper here	Reason for leaving a particular location
5	Give the two main reasons for the growth of megacities.		Migration, including rural–urban migration, and natural increase

PRACTICE

Exam-style questions

1   Explain **one** reason why the development of global communication has contributed to the acceleration of globalisation. **(4)**

2   Explain **one** way in which special economic zones have contributed to the spread of globalisation. **(4)**

3   Explain how **one** indicator or index can be used to measure the degree of globalisation in a particular country. **(4)**

4   Explain **one** reason why many people in developing or emerging economies support the global shift of industry. **(4)**

5   Explain **one** reason why international migration has increased in global hub cities. **(4)**

6   Explain **one** way in which globalisation has increased opportunities for disadvantaged groups. **(4)**

> **EXAM TIP** ◎
>
> For a 4-mark 'Explain' question, your answer should provide a reasoned explanation of how or why something occurs, demonstrating your understanding of the topic through your justification and/or examples.

7   Study **Figure 1**.

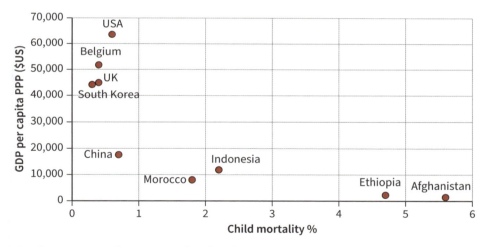

▲ **Figure 1** Gross domestic product (GDP) per capita at purchasing power parity (PPP), and child (under 5 years old) mortality percentage for selected countries in 2021

Exam-style questions

(a) Complete Figure 1 by adding the child mortality and GDP data for Brazil from **Table 1**. **(1)**

Country	Child mortality %	GDP per capita PPP US$
Afghanistan	5.6	1516
USA	0.6	63670
Belgium	0.4	51740
UK	0.4	44979
South Korea	0.3	44232
China	0.7	17603
Ethiopia	4.7	2319
Morocco	1.8	8058
Brazil	1.4	14,592
Indonesia		

(b) Complete Table 1 by adding the child mortality and GDP data for Indonesia from Figure 1. **(1)**

(c) Calculate the range for the child mortality data for the countries shown in Figure 1. **(1)**

(d) State the relationship between child mortality and GDP data shown in Figure 1. **(1)**

8 Explain **one** reason why some groups seek to regain their cultural identity within countries. **(4)**

9 Explain **one** reason why local groups and NGOs promote local sourcing as one response to globalisation. **(4)**

10 Assess the importance of developments in transport as the most significant factor in the acceleration of globalisation. **(12)**

11 Assess the extent to which political and economic decision-making by national governments are significant factors in the acceleration of globalisation. **(12)**

12 Assess the most significant factors that explain why some locations remain largely 'detached'. **(12)**

13 Assess the extent to which global shift has caused more environmental costs than benefits. **(12)**

14 Assess the extent to which migration brings more economic costs than benefits. **(12)**

15 Assess the factors which have led to the emergence of a global culture based predominantly on western ideas and consumption. **(12)**

16 Assess the factors that have led to widening development gap extremities in some countries. **(12)**

17 Assess the attempts made in some locations to control the spread of globalisation. **(12)**

18 Assess the success of local sourcing and ethical and environmental concerns about unsustainability as a response to the spread of globalisation. **(12)**

> **EXAM TIP**
>
> Some 4-mark questions are made up of four separate 1-mark questions. Some of these four questions will assess your quantitative skills. 'Calculate' requires you to produce a numerical answer, showing your workings out. 'Complete' will require you to add detail to a graph or table using additional data that is provided.

◄ **Table 1** Gross domestic product (GDP) per capita at purchasing power parity (PPP) and child (under 5 years old) mortality percentage for selected countries in 2021 data

> **EXAM TIP**
>
> The command word 'Assess' requires you to use evidence to determine the relative significance of something. Your answer needs to be balanced, considering all relevant factors and identifying which is the most important factor, and why.

KNOWLEDGE

5 Regenerating places

5.1 Classifying economies

Classification by sector

- Economies can be classified in different ways and vary from place to place.
- Economic activity can be classified by sector and by type of employment.
- As a country's economy 'matures', it tends to transition from **primary**, via **secondary**, to **tertiary/quaternary** economic sectors.

Many companies have jobs in different sectors, so it can be hard to categorise certain industries. For example, the supermarket Morrisons runs its own farms, manufactures some of its own goods, and employs people in its stores as well as in numerous IT, law, marketing, and other roles.

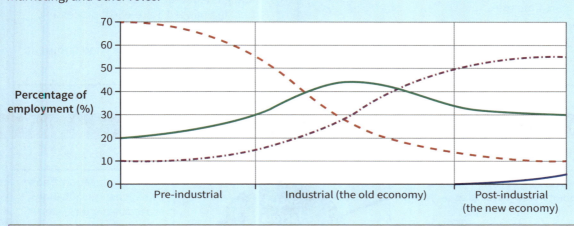

In the UK, the shift from industrial to post-industrial was most noticeable in the 1970s and 1980s.

Key
- - - Primary: extracting raw materials and agricultural produce (e.g. mining, fishing, farming, forestry)
— Secondary: manufacturing, assembling, and processing goods (e.g. steel, cars, processed food)
-·- Tertiary: producing services, either in the private sector (e.g. retail, banking) or public sector (e.g. education)
— Quaternary: providing specialist services in high-tech and knowledge industries (e.g. IT, biotechnology, law)

▲ **Figure 1** *The Clark–Fisher model shows how the four classifications of economies by sector change with time*

Classification by type of employment

Most jobs can be put into the following categories:

|---|
| Part-time (working less than 35 hours per week) or Full-time (working 35 hours or more per week) |
| Temporary (on a restricted length or zero hours contract) or Permanent (on an open-ended contract) |
| Employed (working for a company/organisation) or Self-employed (working for yourself) |

REVISION TIP

Remember that many people now have two or more jobs, as part of the gig economy or as a 'side hustle', so it is difficult to categorise their type of employment. There is also a wide variation in terms of working conditions, pay, and other benefits.

KNOWLEDGE

5 Regenerating places

5.1 Classifying economies

Differences in economic activity

The UK government releases data on economic activity, including:

Employment data

Every three months, information on the number and type of jobs in the UK is released.

National trends show that between 2011 and 2023:

- The number of permanent and full-time jobs has risen.
- Women have taken on more full-time roles and men have taken on more part-time roles.
- The unemployment rate has fallen from 8 to 4%. 22% of people of working age are economically inactive, not available for or seeking work for reasons such as studying, childcare, or ill health. These values are higher in post-industrial cities.

Output data

Every month, an Index of Production for the UK's primary industries and an Index of Services for the UK's secondary, tertiary, and quaternary sector industries are released.

- Whilst coal mining is virtually extinct, oil and gas exploration continues, albeit with fracking paused. Rare-earth processing may begin in some areas.
- Growth industries include wind farms and high-tech industries.

Variations in social factors

Differences in economic activity are reflected through variation in social factors.

Health: There is a clear link between income (amount and reliability) and physical and mental health.

Life expectancy: This is often shorter in deprived areas, due to more dangerous jobs, exposure to air (and other industrial) pollution, stress, and less balanced diets (linked to affordability).

Social factors reflecting differences in economic activity

Levels of education: Those with higher qualifications, especially degrees, are more likely to move to areas with higher-paid jobs, such as cities and the south-east. However, there are huge variations in educational attainment within cities.

Connections between quality of life and economic inequalities

Quality of life indices reflect income inequalities across sectors and types of employment.

- Quality of life is a subjective (opinion-based) degree of social and psychological wellbeing experienced by an individual. Quality of life indices attempt to combine objective (fact-based) measures to enable comparisons.

- Pay tends to be lowest in primary and secondary sector jobs, higher in tertiary sector jobs, and even higher in quaternary sector jobs. However, within all industries in each sector, there will be a hierarchy, with workers 'doing the job' at the

bottom, in lower-paid jobs, and managers and decision-makers at the top, in higher-paid positions.

- Whilst income-deprived neighbourhoods tend to have lower levels of self-reported quality of life, they also tend to score higher on access to some services and on housing affordability.

- In the UK, poorest health tends to occur in the post-industrial areas of south Wales, north-west England, and western Scotland.

 Key terms Make sure you can write a definition for these key terms

employment data   output data
quality of life indices

 REVISION TIP

Practice categorising the economic sectors and types of employment of your local and contrasting places.

5.2 Changing places

Changing functions

Places perform a variety of **functions**, but most places specialise in one or more functions.

Administrative	Retail
Usually cities, where governmental decision-making for the surrounding area takes place.	Some towns and cities are hubs for retail.

Commercial	Industrial
Places with a strong business influence that form bases for many large corporations as well as small and medium enterprises. They are often cities.	Some places are manufacturing hubs, while others have a reputation for high-tech.

The economic basis of places changes over time.

Rural areas in the UK	Urban areas in the UK
• Many rural areas have shifted from predominantly production (i.e. farming, forestry, and mining) to post-production (e.g. leisure and tourism) – a sectoral shift from primary to tertiary. • Compared to farming, jobs in the leisure and tourism industry are more likely to be part-time and temporary. • Some rural areas have evolved into dormitory settlements where residents commute elsewhere to work. • Others have lost residents as properties have been bought as second homes or to rent to holidaymakers.	• Many cities have seen the loss of manufacturing activity to factories abroad – this global shift since the 1970s led to deindustrialisation, which particularly hit cities based on heavy industry (e.g. steel) or textiles (e.g. wool). • Automation and other high-tech changes have also led to the loss of jobs and sectoral shift. • Some jobs replacing those that have been lost are temporary or lower paid, e.g. in call centres, food and drink outlets, warehouses or with delivery firms. • Others have been permanent and higher paid, e.g. some in the finance and business service (FBS) sector.

Changing demographic characteristics

The **demographic** characteristics of a place change as its function changes.

Gentrification

The social status of a place increases due to the movement of higher socio-economic groups to inner-city areas. This often follows deindustrialisation, and the shift towards a service-based economy. **Gentrification** can be uncoordinated or planned. It is often resisted by local residents.

Age structure

Urban areas in economic decline will often have an older age profile, as young adults of working age leave to find jobs. Regenerated areas often attract young adults. Areas with a high proportion of recent migrants will tend to have more young adults, especially males, and more children.

Composition

Over time, inward migration might change an area's population and **ethnic mix**, and existing residents might leave. Newcomers may be attracted by jobs or **regeneration** projects. In recent decades, gentrification has sometimes resulted in the displacement of migrant communities.

> **REVISION TIP**
>
> Ensure you understand that gentrification now means any way in which areas are changing to attract people from a higher socio-economic status.

KNOWLEDGE

5 Regenerating places

5.2 Changing places

Reasons for change

Physical factors

Some physical factors that were crucial for the development of industries, such as rivers to supply water for textile industries, or coalfields for steel manufacture, are no longer influential for most places in the UK. Instead, the threat of river and coastal flooding has necessitated the construction of defences and re-zoning.

Historical development

Places that grew before the advent of the car tend to have narrower, unplanned street patterns, which are inconvenient for manufacturing industries, but attractive to shoppers and tourists. Out-of-town retail and science parks are planned around road access.

Accessibility and connectedness

Access to train lines and motorways can help places to attract residents and inward investment. Proximity to regional and international airports may also help places whose economies are based around tourism, or who have strong international business sectors.

Local and national planning

Planners will zone areas to encourage certain functions to grow. National governments previously built new towns such as Cumbernauld in Scotland to relieve pressure on densely populated cities. In the 1980s, the UK government set up urban development corporations (UDCs) to regenerate post-industrial 'brownfield' land in inner-city areas such as London Docklands. National infrastructure projects like HS2 can attract development.

REVISION TIP

Find out how and why the functions and demography of your local and contrasting places have changed in recent decades.

Measuring change

- Employment trends: unemployed versus employed, sectoral shift, graduate jobs, level of pay, household income. Sources of data: census, ONS surveys.

- Demographic trends: age profile, ethnic diversity, male/female balance. Sources of data: census, GP Fingertips profiles, ONS surveys.

- Land-use changes: residential, commercial, retail, industrial, derelict, green space, blue space (water). Sources of data: OS maps, primary data (land-use surveys).

- Levels of **deprivation**: primary deprivation data on income, employment, health, crime, quality of living environment, and abandoned and derelict land.

The Index of Multiple Deprivation (IMD) provides a set of relative measures of deprivation for small areas across England.

Sources of data:

- income deprivation and employment deprivation – IMD

- health deprivation – self-reported health (census), IMD

- crime – British Crime Survey (victims), police.uk (reported crimes), IMD, primary data

- quality of the living environment – IMD, primary data

- abandoned and derelict land – land-use survey, satellite imagery, primary data, IMD (the IMD also measures barriers to housing and services).

Key terms Make sure you can write a definition for these key terms

administrative   commercial   demographic   deprivation
ethnic mix   function   gentrification   regeneration

5.3 Influences on changing places

Global and international influences

Places change function and characteristics over time because they are affected by international and global influences.

Global and international influences

TNCs:
Global brands with functions or parts of their business in different countries to exploit the resources there. They create jobs.

Deindustrialisation:
TNCs are able to move production to other countries.

Competition:
Much of the world is one big market due to globalisation, increasing competition for resources and customers.

Conflict:
War and civil war can impact on the availability of products and raw materials, levels of immigration, and trade between countries.

Tourism:
The unique features of some places are very attractive to international visitors.

Political and economic groups:
Being part of such a group can provide access to trade agreements and development funds.

Transport and internet connections:
Locations have variable access to superfast broadband networks and fast transport links.

National and regional influences

Places change function and characteristics over time because they are also affected by regional and national influences.

National and regional influences

Physical (geographical) characteristics:
Proximity to fresh water and to rivers or coastline; remoteness or accessibility due to the terrain.

Transport connections:
Proximity to major transport route intersections, rail connections, airports, etc.

Institutional connections:
Major institutions (academic or commercial) working collaboratively.

Industry:
Long associations with certain industries due to the availability of resources, including skilled labour.

National policies:
Immigration policies and initiatives to boost economic growth.

Local organisations:
The activity of tourist boards, community groups, and charitable trusts can boost a region's profile.

 **KNOWLEDGE**

5 Regenerating places

5.3 Influences on changing places

Characteristics of your chosen places

Global influences

- Has the global shift of manufacturing and certain services, especially to Asia, affected your places?
- Have global population movements, conflicts, or political, cultural, or environmental trends affected your places?

International influences

- What TNCs have headquarters or other operations in your places?
- Have your places received funding or investment from the EU or other IGOs?
- Are there any destinations which attract international tourists and/or students in your places?

National influences

- What national businesses have headquarters or other operations in your places?
- What national government initiatives have taken place in your places?

Regional influences

- What local businesses have their headquarters in your places?
- What local government initiatives have taken place in your places?

REVISION TIP

Summarise the characteristics of your chosen places, covering each of these points. Consider the demographics, culture, land use, and built and natural environments of the places.

REVISION TIP

Find images, maps, and graphs that represent your chosen areas, and make annotated collages of both areas.

Places can be represented in different ways

- What have various groups lost (**cultural erosion**) or gained (**cultural enrichment**)?
- How have ways of life changed? Think about religion, language, food, drink, dress, customs, etc.
- How do various groups of people feel about the changes?

REVISION TIP

Remember that your chosen and contrasting places should be a locality, a neighbourhood, or a small community, *not* an entire city.

How the lives of students and others are affected

- How are the lives of you and other students affected by continuity and change in your places?
  - Where do students spend any available free time?
  - Where do students work?
  - How do students get around?
- How are the lives of other people affected by continuity and change in your places?
  - Where do other people spend any available free time?
  - Where do other people work?
  - How do other people get around?

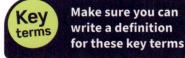 **Key terms** Make sure you can write a definition for these key terms

cultural enrichment   cultural erosion

⇄ RETRIEVAL

Learn the answers to the questions below, then cover the answers column with a piece of paper and write as many as you can. Check and repeat.

Questions

Answers

#	Question	Answer
1	Name the economic sector that provides services.	Tertiary
2	Name the model that shows how the percentage of people employed in each economic sector changes over time.	Clark-Fisher model
3	Which era saw the peak of employment in primary industries in the UK?	Pre-industrial
4	In which sector of employment would IT and biotechnology be classified?	Quaternary
5	What are the two main types of data that show economic activity?	Employment and output
6	Name the three social factors that reflect differences in economic activity.	Health, life expectancy, and levels of education
7	What function relates to governmental decision-making?	Administrative
8	What do places with commercial functions provide?	Business influence and activity
9	What are dormitory settlements?	Places where residents commute elsewhere to work
10	What does 'demographic' mean?	Related to people (especially age, gender, and ethnicity)
11	Which process describes the movement of higher socio-economic groups to inner city areas?	Gentrification
12	What happens to the age structure and ethnic composition of predominantly migrant neighbourhoods?	Age falls and ethnic composition can become more diverse
13	Name two ways that national planning decisions have led to places changing.	Two from: zone areas / new towns / urban development corporations (UDCs) / national infrastructure projects
14	What is an IMD and what does it measure?	Index of Multiple Deprivation; deprivation
15	Would the loss of a community centre be cultural erosion or cultural enrichment?	Cultural erosion

Put paper here

⚙ KNOWLEDGE

5 Regenerating places

5.4 Social and economic inequalities

Successful places

High rates of employment, especially well-paid and highly skilled jobs, usually in the tertiary and quaternary sectors.

High rates of inward migration.
- Internal migration:
  — Rural hinterland (area surrounding the place).
  — Smaller urban areas in the region, which offer fewer opportunities.
  — Declining urban areas in the rest of country.
- External migration from outside the country.

Signs that a place is 'successful'

Low levels of **multiple deprivation**.
- High income
- High levels of education, skills, and training
- High levels of long-term health
- Low levels of crime
- Ease of access to services
- High quality living environment
- Access to employment

REVISION TIP

Remember, the idea of a 'successful' place is a subjective one, meaning that it depends on whose opinion is being sought.

Negative effects of success

- High demand for housing leads to high purchase prices, high rents, and a shortage of **social housing**.
- High retail prices, as shops seek to maximise their intake.
- Rapidly changing cultures, as incomers may bring different ways of life.
- Skill shortages, especially in jobs with modest salaries, as these workers are not able to afford housing, so they may move away from the area.

Positive and negative effects occur in both urban and rural areas

Successful urban area: Vienna, Austria

Named by *The Economist* in 2023 as the world's most liveable city, Vienna offers:
- ☑ a range of well-paid jobs (e.g. banking, government, and specialist manufactured goods)
- ☑ political stability and membership of EU (trade bloc)
- ☑ good infrastructure (e.g. trams and underground)
- ☑ strong education (five universities) and healthcare (40 hospitals)
- ☑ culture and entertainment (especially music).
- ☒ However, housing demand and costs are very high.

Successful rural area: Cheshire's Golden Triangle, England

- ☑ An area of small towns and villages between Wilmslow, Alderley Edge, and Prestbury.
- ☑ Pleasant countryside – clean air, many green spaces.
- ☑ Easy access to Manchester.
- ☑ Near to motorways (M6 and M62) and Manchester Airport.
- ☑ Access to good schools and health facilities in Manchester.
- ☒ However, many of the original residents and their descendants have been priced out of the area, and shops and restaurants are expensive.

Effects of economic restructuring

In some regions, **economic restructuring** has triggered a spiral of decline in urban and rural areas. Sometimes an entire sector can shrink, e.g. the automotive sector in Detroit, USA. This results in deindustrialisation, which triggers a **spiral of decline** which can be difficult to escape. This can take place in urban areas, but also in rural ones (often based on primary industries, e.g. coal mining).

The Rust Belt, NE USA

Until the 1950s, the area from Pennsylvania to Iowa was the world's biggest heavy industrial region.

Deindustrialisation ocurred, due to global shift, mechanisation and lower wages in SE USA and Mexico.

High-income jobs in the primary and secondary sectors have been replaced by low-wage tertiary jobs in retail and local government.

REVISION TIP

Make sketches to show urban processes. For example, draw a spiral of decline or a systems diagram (inputs, processes, and outputs).

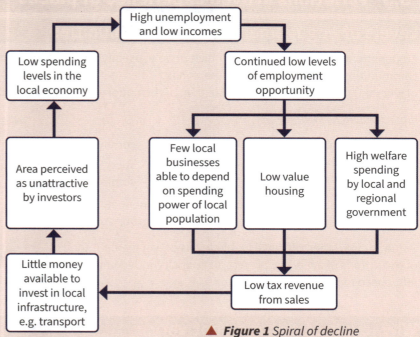

▲ *Figure 1* Spiral of decline

Economic and social inequalities in urban and rural locations

Gated communities	Poor quality housing
High-cost housing situated behind a barrier with monitored access. There may be rules that lead to uniform appearance of gardens and exterior designs. Outsiders can feel observed and unwelcome, causing social segregation and suspicion.	When local authority (council) tenants were granted the right to buy in 1980, many bought the best quality houses in the most attractive neighbourhoods. The remaining tenants (often vulnerable people) were concentrated into the worst quality housing stock. National governments now provide little funding for the construction and maintenance of social housing, leading to a housing shortage.
Commuter (dormitory) villages	**Declining rural settlements**
Usually on the outskirts of economically successful cities, these have limited weekday retail and cultural life. High concentration of well-paid workers and families but fewer older and lower-income residents, leading to social segregation and suspicion.	Rural employment opportunities (e.g. farmwork, mining) lost through mechanisation, import of minerals, and reduced use of coal. Second-home ownership and holiday cottage rentals cause higher prices, so local residents are displaced. Young adults hope to leave for education, jobs, and social opportunities.

Key terms Make sure you can write a definition for these key terms

economic restructuring   gated communities
multiple deprivation   social housing   spiral of decline

5 Regenerating places

5.5 Variations in lived experience of places

Levels of engagement in local communities

Engagement refers to the ways in which people, usually residents, interact with local communities. It usually refers to in-person activities (e.g. attending sports clubs or places of worship), but it can also be remote (e.g. participation in online community forums).

It can involve:

- political engagement
- electoral turnout (both national and local)
- community engagement
- sport, places of worship, other community groups.

Many community groups have grown as government austerity measures have led to councils removing funding (e.g. libraries, youth groups).

Some communities will have higher levels of engagement than others – often because of the actions of a few key players, or council-funded initiatives. It may also be influenced by socio-economic factors.

> **REVISION TIP**
>
> Search online for groups in your area and note down three to five activities that they run which involve the community (e.g. fairs, community gardening, litter picks, guided walks).

Why levels of engagement vary

Everyone's **lived experience** of a place is unique and will affect their **attachment to a place** and level of engagement with it. Lived experience varies due to:

Age

- School-age children rely on how their parents engage, but tend to have a small, walkable, sphere.
- Working age residents split their place attachment between home, work, and hobbies; they usually engage with a wider sphere.
- Retired people tend to have a deeper place attachment; they often have time to engage and are more likely to have restricted/limited mobility.

Length of residence

- University students may not develop deep roots in an area – they usually stay for only one to five years.
- New migrants may rely heavily on community facilities; levels of integration may vary.

Levels of deprivation

- Affluent residents can join clubs and spend money locally, but spend time away too.
- More deprived residents may rely on neighbour and family support.
- Unemployed or non-economically active residents may shop and volunteer more locally.

Ethnicity

- Ethnic groups in the majority can access more community facilities and shops.
- Ethnic groups in the minority may experience language and cultural barriers to experiencing places.
- Over time, the above factors are reduced as groups integrate.

Gender

- Traditionally, men may engage less (more links elsewhere, e.g. work) but may have loyalty (e.g. support sports team).
- Traditionally, women may engage more (if their main responsibility is raising a family, e.g. school, clubs, etc.).
- However, gender roles are becoming more equal in families and employment, and the relationship between gender and lived experience is complex.

> **REVISION TIP**
>
> Remember that these factors overlap and are dynamic (they change over time); societal trends and attitudes also change. Take care to delve deeper into place attachment and engagement, ideally referring to your experiences or those of people you have talked to.

Conflicts around regeneration

Conflicts can occur between groups over what should be prioritised for regeneration and what strategies could be used.

Does the area need social, economic, or environmental assistance?

- Small and medium-sized enterprises (SMEs) may want help with tax breaks, energy costs, infrastructure.
- Some residents may want anti social behaviour to be targeted.
- Health campaigners may want more affordable healthy food and sports facilities.

Who decides?

- Top-down schemes (from local or national governments) tend to be better funded but may not be flexible enough to respond to concerns from locals.
- Bottom-up schemes (from local groups) may represent the voice of one local group, but perhaps not others. Also, the local group may be poorly funded and dependent on one key personality.

Who delivers the regeneration?

- Local authorities may listen to the community or just part of it; bureaucracy can take time.
- Private companies may be incentivised to complete projects quickly, but their focus on profits might sideline social aspects. (For example: do housing estates have playing areas, shops, or clinics?)

Other causes of conflict

- Property-led regeneration may attract new, wealthier groups, whose values and lifestyles might clash with those of pre-existing residents.
- Political engagement may be low: time-poor and less confident residents may not be able to attend, or feel able to speak out, at planning meetings.
- Some groups may be perceived as holding power, where it is felt that the needs of some groups are prioritised over the needs of others.
- Many developments may boast their job-creation potential, but jobs created may provide little economic opportunity, especially for residents with lower levels of education – but skills take time to build up.
- Long-term priorities may be important (economic restructuring, air quality), but some residents have more pressing immediate needs (cost of living).
- Low traffic neighbourhoods (LTNs) have led to conflicts, as environmental priorities (lowering air pollution and accidents) clash with the need for some residents to conduct their businesses or visit relatives. Similar conflicts have arisen with ultra-low emissions zones (ULEZs).

REVISION TIP

Use your own experience of places from your home and school, and research from your independent investigation.

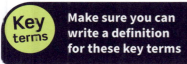

Key terms Make sure you can write a definition for these key terms

attachment to place   conflicts   engagement   lived experience

5 Regenerating places

5.6 Evaluating the need for regeneration

Using statistical evidence to determine the need for regeneration

- Statistics can give evidence of the need for regeneration in a place.
- Two key sources of statistical evidence for regeneration are the 2021 census and the IMD (for England only).

Census data	IMD
- Each census provides information about the population structure (e.g. age, gender, ethnicity, employment, education, etc.) that can be used by local authorities. - Data reveal who is missing from the population structure (e.g. data may show a lack of young professionals). This helps authorities identify the demographic they want to attract through regeneration. - Data will also inform authorities which people in the population structure are likely to be most affected by any regeneration scheme.	- As a relative measure of deprivation, the IMD provides data against seven indices (criteria): income, employment, education, skills and training, health, crime, housing, and living environment. - Data allow comparison of different small areas nationally, regionally, or within a local authority, so areas requiring funding for regeneration can be identified and appropriate schemes can be devised to target particular types of deprivation.

Data for your chosen places

Category	Statistical evidence to consider
Economic	Census: - % of residents in full-time work IMD decile for: - income deprivation - exclusion from employment
Social	Census: - % of residents with no educational qualifications - % of households renting or owning property - % of households with high bedroom occupancy ratings IMD decile for: - deprivation of education, skills, and training - health deprivation and disability - barriers to housing and services - crime
Environmental	IMD decile for living environment deprivation Green space: magic.defra.gov.uk Air pollution: naei.beis.gov.uk (emissions map)
Composite	IMD: What rank is your Lower layer Super Output Area (LSOA) on the list of 32,844 LSOAs? What decile (10% band) does this place your area in?

REVISION TIP

Research and analyse data for both your chosen places, considering all the points listed in the table.

Media presentations of places

Different media can provide contrasting evidence, questioning the need for regeneration in your chosen places.

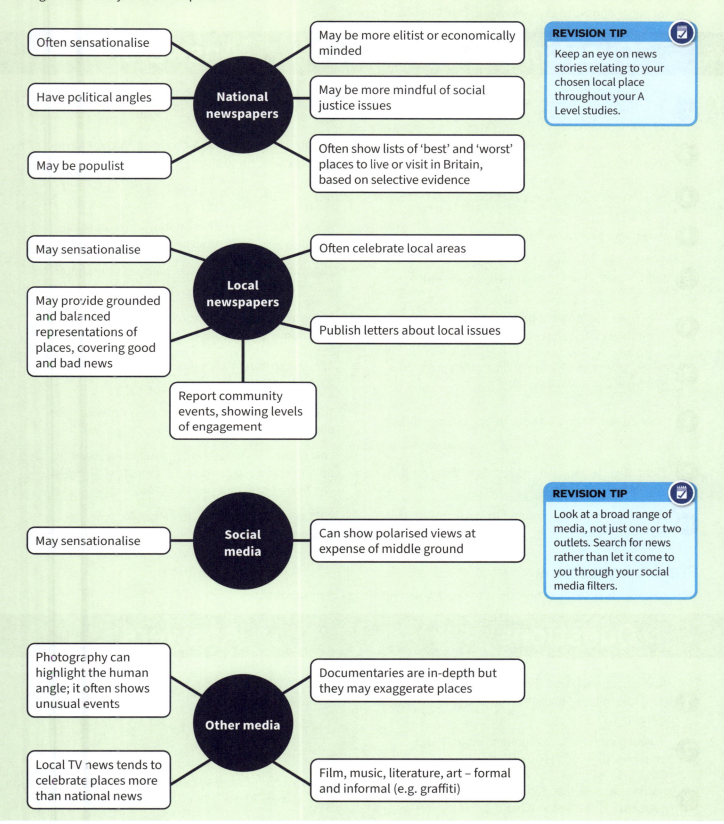

Often sensationalise

Have political angles

May be populist

National newspapers

May be more elitist or economically minded

May be more mindful of social justice issues

Often show lists of 'best' and 'worst' places to live or visit in Britain, based on selective evidence

May sensationalise

May provide grounded and balanced representations of places, covering good and bad news

Local newspapers

Often celebrate local areas

Publish letters about local issues

Report community events, showing levels of engagement

May sensationalise

Social media

Can show polarised views at expense of middle ground

Photography can highlight the human angle; it often shows unusual events

Local TV news tends to celebrate places more than national news

Other media

Documentaries are in-depth but they may exaggerate places

Film, music, literature, art – formal and informal (e.g. graffiti)

REVISION TIP

Keep an eye on news stories relating to your chosen local place throughout your A Level studies.

REVISION TIP

Look at a broad range of media, not just one or two outlets. Search for news rather than let it come to you through your social media filters.

RETRIEVAL

Learn the answers to the questions below, then cover the answers column with a piece of paper and write as many as you can. Check and repeat.

Questions

Answers

	Questions	Answers
1	Name the three general signs that a place can be considered 'successful'.	High employment, high inward migration, low deprivation
2	List two possibly negative effects of 'successful' regeneration.	Two from: high demand for housing / high retail prices / rapidly changing cultures / skill shortages
3	What might economic restructuring trigger?	A spiral of decline
4	Why might gated communities be problematic?	Social uniformity (within estate), and deepening social segregation and suspicions of others
5	Write a definition for the term 'engagement'.	The ways in which people, usually residents, interact with local communities
6	Give three examples of political or community engagement.	Three from: voting / sports clubs / places of worship / community groups
7	What five key factors can make everyone's lived experience of a place unique?	Age, length of residence, levels of deprivation, ethnicity, gender
8	How do top-down and bottom-up schemes differ?	Top-down schemes come from governments; bottom-up schemes come from local groups
9	What are LTNs, and why have they caused conflicts?	Low traffic neighbourhoods; benefits are lowering air pollution and accidents, but they restrict the freedom and flexibility of businesses and residents
10	What are the two key sources of statistical evidence for regeneration?	2021 census and the Index of Multiple Deprivation (IMD)

Put paper here

Previous questions

Now go back and use these questions to check your knowledge of previous topics.

Questions

Answers

	Questions	Answers
1	Name the model that shows how the percentage of people employed in each economic sector changes over time.	Clark-Fisher model
2	In which sector of employment would IT and biotechnology be classified?	Quaternary
3	Name two ways that national planning decisions have led to places changing.	Two from: zoned areas / new towns / urban development corporations (UDCs) / national infrastructure projects
4	What is an IMD and what does it measure?	Index of Multiple Deprivation; deprivation
5	What do places with commercial functions provide?	Business influence and activity

Put paper here

5 Regenerating places

5.7 UK government policies for regeneration

Infrastructure investment

Infrastructure	Social infrastructure
Transport	Housing
Energy distribution	Clinics and hospitals
Water and sewerage facilities	Schools and universities
Telecommunication networks	Youth clubs and community hubs

Infrastructure investment can maintain economic growth and improve accessibility to regenerate regions by:

- speeding up the flow of raw materials and goods
- increasing the commuter threshold (the distance commuters are willing to travel to work)
- attracting **foreign direct investment (FDI)** as companies seek access to new markets.

High Speed 2 (HS2)

- A transport infrastructure investment, originally supposed to link London, Birmingham, Manchester, Sheffield, and Leeds.
- Approved in 2012 (budget £33bn); work began in 2017 (budget rose to £106bn before the Yorkshire leg was cut).

	Costs	Benefits
Economic	• High economic cost (money could be spent elsewhere – the opportunity cost)	• Enhanced capacity; people can work on the train • Attracting FDI and stimulating growth
Social	• Displacement and disruption to places along the route • Loss of culture, e.g. in the area around Euston	• Boosts connectivity for tourists and residents • Job creation in construction and near the stations
Environmental	• Destruction of natural habitats, e.g. in the Chilterns • Urban sprawl	• Displaces domestic air travel, so fewer CO_2 emissions • Less car travel so fewer CO_2 emissions, particulates, etc.

5 Regenerating places

5.7 UK government policies for regeneration

Airport development

- Airport development allows rapid travel within and between UK and European cities, and boosts the UK tourist industry. Business travellers may be more likely to invest in the UK.

- Airport infrastructure may be locally disruptive (e.g. loss of habitat, noise pollution, traffic), but it avoids the problems along routes involved in train and car travel.

- CO_2 emissions are the main environmental drawback, despite development of lower-carbon planes and fuels.

- The UK government partners with developers (e.g. airport companies like Manchester Airports Group) and charities (the Big Lottery Fund supports local and regional projects like the 2018 Great Exhibition of the North).

Housing development policies

The rate and type of development affects economic regeneration. When considering which development to support, the government may prioritise national over local needs and opinions.

> **REVISION TIP**
>
> You need to remember the links between planning laws, house building targets, and housing affordability.

Planning laws

Attempt to limit the negative impact of unregulated development. Without them, property developers would be tempted to follow the profit imperative and build larger family homes, in greenfield areas (sprawl).

But relaxing planning laws can lead to economic growth as more workers could move to the development area.

House building targets

Popular on a national level as they increase housing supply and reduce costs. Unpopular at a local level, as people lose views and green space. Retrofitting old housing would be more eco-friendly.

Targets have been missed since 2007. In 2019, the government set a target of 300,000 a year by the mid-2020s. The closest to this was 242,700 in 2019–20.

Over 95% of housing in the 2010s was private; local authority housing fell rapidly in the 1980s; housing associations provide some housing.

Housing affordability

Linked to planning laws and targets. More available houses should lead to lower prices. But smaller households, separated families, longer life expectancy, and high net immigration means demand is rising faster than supply.

The UK government has a Help to Buy scheme (purchasers borrow 20% to 40% of the price), permits shared mortgages, and makes developers build 'affordable housing'.

But there is no price control on rents, and the main political parties do not usually want to upset homeowners or landlords.

Government policy on international migration

Decisions about **international migration** can promote or restrict growth and investment.

Open-door policy

When the UK was part of the EU, it respected free movement of labour from member states. Many businesses grew as EU workers filled skills gaps, often working for lower wages. Tax revenues increased, providing, for example, the potential funds to support more pensioners. But many pre-existing residents were displaced from jobs or saw wage stagnation.

Closed-door policy

In its preparations for Brexit in the late 2010s, many in the UK government wanted to slow the flow of immigrants and develop UK workers' skills, boost wages, and encourage the flow of money in the economy.

Since the 2010s, asylum seekers have been discouraged by detention and threat of deportation. Morality aside, the UK government's economic argument is that looking after refugees is costly to the economy, despite many having useful skills.

Targeted policy

In 2020, the UK government introduced a points-based immigration system. It allows migrants with a skill that is in short supply here, and those who have an employer sponsor, to enter the UK. Some university students and others are also allowed into the UK. Targeting skills shortages can help the UK economy increase productivity. The education sector gains fees from international students, and some students may stay and become tax-paying workers in the UK.

> **REVISION TIP**
>
> Keep an eye on the news for stories relating to UK government policies throughout your A Level studies.

Financial deregulation

Financial deregulation can promote growth and investment.

Changes in 1986

Anyone can trade shares (not just stock exchange) → More funds for companies to invest → Companies expanded → Companies employed more people

Overseas financial institutions can set up in UK → Finance and business services accounted for 8.3% of UK GDP in 2021 → Property sector boomed from increased demand for office space (e.g. London Docklands) → Construction sector grew, and its employees earned money which increased the amount circulating around the economy

Disadvantages of financial deregulation

✗ Financial deregulation has opened property and infrastructure (e.g. water and energy) to overseas investment, and allowed speculation, resulting in booms and busts, e.g. 2008/9 financial crisis. Property booms and busts also occur.

✗ With the increase in working from home (WFH), overseas banks can operate from abroad yet still influence the UK economy.

Key terms Make sure you can write a definition for these key terms

closed-door policy   foreign direct investment (FDI)
financial deregulation   infrastructure   international migration
open-door policy

5 Regenerating places

5.8 Local government policies

Role of local government in regeneration

- Structural activities, e.g. investing in physical infrastructure and social infrastructure, provide the basis for the successful operation of a district.
- Strategic activities attempt to attract investment, residents, or visitors. They tend to target one or more areas within a district, and they are usually time-limited.

Local governments compete to attract businesses to increase their tax base and pool of skilled labour, and to attract linked industries. Much of the tax yield can then be spent on infrastructure, which attracts more industries.

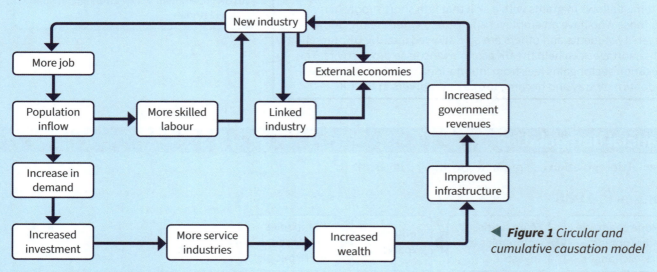

◀ **Figure 1** Circular and cumulative causation model

Science and technology parks

This model is most relevant to secondary industries, which are **footloose** (they do not need to be located near raw materials or large pools of labour). Footloose industries include many science and technology firms.

Footloose firms will have a set of **location factors**. Local governments have responded by developing science and technology parks that offer:

- Proximity to skilled labour (usually graduates)
- Close links to universities (to take advantage of new research)

- Good business infrastructure (broadband and energy)
- High-tech premises that are aesthetically pleasing and may have simplified planning laws and subsidised rents
- Good transport links (road, rail, air) for workers, suppliers, and access to market
- Proximity to suppliers and clients (for ease of meeting and business synergy).

Enterprise Zones

Regional governments have developed larger **Enterprise Zones**, which leverage national government funding across a large area to offer business rate discounts, tax relief on machinery, and simplified local planning. One example is the Leeds City Region M62 Corridor Enterprise Zone.

Local interest groups

Local interest groups have different priorities:

- Chambers of Commerce want to maximise access to customers and minimise planning restrictions. They often prioritise existing members over newcomers.

- Local preservation societies want to use planning restrictions to curb developments they feel may drastically change the heritage value, aesthetics, and/or environmental benefits of an area.

- Trade unions will usually welcome new investments, subject to pay and working conditions for workers.

- Developers want to maximise profits and minimise the impact of regulation, e.g. building the minimal number of affordable housing and leaving the smallest possible green spaces.

Local governments rely on local and national taxation to fund their activities. They need to work within legal and financial limits set down by the national government. They sometimes also seek funding from the private sector for what is known as **partnership-led** regeneration. Tensions may exist between these groups.

> **REVISION TIP**
>
> Avoid referring to 'local residents', and instead consider which groups of residents hold which opinions.

Urban and rural regeneration

Urban and rural regeneration strategies can be:

- Skills- and business-led, e.g. York Science Park

- Property-led, e.g. Canary Wharf, London

- Retail-led, e.g. the Meadowhall shopping centre on post-industrial land in Sheffield

- Tourism-led, e.g. branding Haworth in West Yorkshire as 'Brontë Country'

- Sports-led, e.g. Queen Elizabeth Olympic Park, London

- **Diversification**, e.g. Grow in Powys, Wales

- Culture-led, e.g. the UK City of Culture scheme

Tourism, leisure, sport, and retail-led plans: Queen Elizabeth Olympic Park, London

The 2012 Olympic Park in East London reopened in 2014 as the Queen Elizabeth Olympic Park.

✅ Developers, the UK government, the London Mayor, and Newham Council supported the Olympics and the post-Olympic legacy. It transformed post-industrial derelict land into sporting venues, 2800 flats and apartments, Stratford's Westfield Shopping Centre, and Stratford International and regional train stations.

✅ Environmental groups were generally pleased with the 'soil hospital' which decontaminated the land and attempts to encourage the close provision of live–work units, encouraging active travel.

❌ Many residents of local authority housing, Traveller communities, and several active industries were opposed to the plans, as they involved the compulsory purchase of existing buildings and the displacement of residents. Many lower-income residents were priced out of the housing market, and their local community was lost.

Developments continue with the opening of the East Bank cultural quarter in 2022, featuring hubs for the BBC, University College London, the V&A Museum, and other institutions. House prices remain high.

Public/private rural diversification: Grow in Powys, Wales

- Powys Council wants to exploit the extensive and beautiful landscape of central Wales to encourage a diverse economic base. It advises and funds local businesses, social enterprises, and community organisations in green tourism, creative industries, carbon storage, community broadband, local food production, renewable energy, water regulation, and town centre loans.

- The strategy aims to help the county to escape its low-wage economy and reverse its skills shortage in a **sustainable** manner.

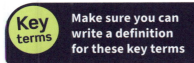

Key terms Make sure you can write a definition for these key terms

diversification   Enterprise Zones   footloose
location factor   partnership-led   sustainable

5 Regenerating places

5.9 Rebranding

Improving the image of rural and urban places

Rebranding involves **re-imaging** places to improve their image (highlighting existing ideas or creating new ones), make them stand out, and make them more attractive for investors. This can also involve physical change (new buildings, infrastructure etc.).

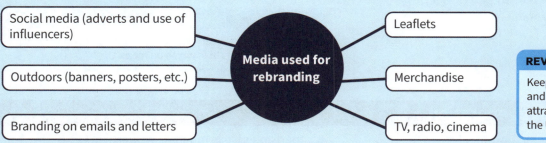

Social media (adverts and use of influencers)

Outdoors (banners, posters, etc.)

Branding on emails and letters

Media used for rebranding

Leaflets

Merchandise

TV, radio, cinema

 REVISION TIP

Keep an eye out for adverts and publicity aimed at attracting you to places in the UK.

Rebranding deindustrialised UK cities

Rebranding can emphasise the attraction of deindustrialised UK cities, using their industrial **heritage** to attract national and international tourists and visitors.

	Belfast	Bradford
Original key industry	Shipbuilding	Textiles (especially wool)
Rebranding	Titanic Quarter	2025 City of Culture
Key buildings	• Titanic Belfast visitor centre • Land of the Giants arts venue	• National Science and Media Museum • City Park • Bradford Live (new music venue)
Key events	• BBC Proms in the Park • Maritime Festival	• 1000 performances • 365 artist commissions
How brand builds on previous industry	• Huge cranes generate awe and wonder • Belfast Harbour continues to operate • Maritime Mile initiative encourages people to walk/cycle along riverside	City of the World theme will look back at Bradford's time as a world leader in industry, as well as the people from various cultures that have made Bradford their home

Assessing urban rebranding

Advantages	Disadvantages
☑ Creates jobs and skills (e.g. Bradford 2025 claims 4500 direct jobs, 2500 indirect jobs, 4000 volunteers). ☑ Creates an income stream. ☑ Gives residents a sense of pride of place.	☒ Diverts funds from social infrastructure and more immediate needs. ☒ Multiple deprivation is difficult to improve and has structural causes related to the national economy. ☒ Rebranding is often focused on tourism, which can be seasonal and fickle. ☒ Cultural events may make areas 'cool', but affluent incomers may displace locals and erode cultures.

Rural rebranding strategies

- Agriculturally, the UK countryside is still very productive, but greater efficiencies and food imports mean less land is used for farming and there is more post-production countryside.
- Squeezed profit margins, depopulation, and an ageing population also create challenges for rural areas.
- Rural rebranding strategies are needed.

Heritage and literary associations

For example, Brontë Country:

- Bradford Council exploits its link to the Brontë sisters (including the visitor centre) in its publicity.
- The Keighley and Worth Valley steam railway and cobbled main street have heritage value.
- English Heritage and Bradford Council issue grants for 'staged authenticity'. This keeps historicity visible.
- Hosts 1940s and 1960s themed weekends.
- Most visitors come for a day. International tourists often visit then stay overnight elsewhere.

Farm diversification and specialised products

For example, Organic Pantry, Newton Kyme, North Yorkshire (onsite shop, box delivery, farmers markets).

- Around 62% of UK farmers diversify alongside running a working farm. For example:
  - camping, glamping, farm stays, sports (e.g. four-wheel drive activities, water-based recreation)
  - solar and wind farms
  - specialised products (e.g. artisan cheeses and organic crops).

Outdoor pursuits and adventure

For example, Kielder Forest, Northumberland – 'The greater outdoors':

- Dark skies (the darkest in England) are publicised to attract astronomers and naturalists.
- Activities for naturalists, e.g. Tower Knowe Visitor Centre, trails, online guides, and apps.
- Outdoor activities, including Kielder Marathon, ultras, triathlons, sailing.

Assessing rural rebranding

Advantages	Disadvantages
☑ Reduces rural depopulation.	☒ Often focused on tourism, which can be seasonal and fickle. Locals may feel they are viewed like a tourist attraction.
☑ Balances age structure.	☒ Can lead to transport congestion.
☑ Creates jobs, skills, and income.	☒ High-tech industries may not see 'preserved' places as being open to investment.
☑ Pride of place.	

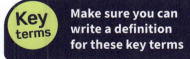

Key terms Make sure you can write a definition for these key terms

heritage   rebranding   re-imaging

RETRIEVAL

Learn the answers to the questions below, then cover the answers column with a piece of paper and write as many as you can. Check and repeat.

Questions

		Answers
1	Give three examples of infrastructure.	Three from: transport / energy distribution / water and sewerage facilities / telecommunication networks
2	Name the three broad types of international immigration policy.	Open doors, closed doors, targeted
3	Name four location factors for science and technology firms.	Four from: transport links / high-tech premises / infrastructure / proximity to universities / proximity to skilled labour / proximity to suppliers and clients
4	What are two benefits of Enterprise Zones for businesses?	Two from: business rate discounts / tax relief on machinery / simplified local planning
5	What is a partnership-led regeneration strategy?	One which relies on funding from public and private sources
6	Name two players who were generally opposed to the Queen Elizabeth Olympic Park plans.	Two from: residents of local authority housing / Traveller communities / several active industries
7	Name three ways in which Grow in Powys could be said to be an environmentally sustainable regeneration scheme.	Three from: green tourism / local food production / renewable energy / carbon storage / water regulation
8	Give two ways in which deindustrialised urban areas can use their heritage to rebrand.	Two from: museums / visitor centres / awesome structures / use of derelict land to host buildings or events

Put paper here (between columns)

Previous questions

Now go back and use these questions to check your knowledge of previous topics.

Questions

		Answers
1	List two possibly negative effects of 'successful' regeneration.	Two from: high demand for housing / high retail prices / rapidly changing cultures / skill shortages
2	Write a definition for the term 'engagement'.	The ways in which people, usually residents, interact with local communities.
3	What five key factors can make everyone's lived experience of a place unique?	Age, length of residence, levels of deprivation, ethnicity, gender
4	How do top-down and bottom-up schemes differ?	Top-down schemes come from governments; bottom-up schemes come from local groups
5	What are the two key sources of statistical evidence for regeneration?	2021 Census and the Index of Multiple Deprivation

Put paper here (between columns)

 **KNOWLEDGE**

5 Regenerating places

5.10 Assessing the success of regeneration

Measures of income, poverty and employment

Income

Income can be measured using the IMD dimension of income deprivation, showing if people are out of work or on low earnings. Household income includes wages and income from property and financial assets.

> **REVISION TIP**
>
> Remember that national patterns in income hide wide local differences.

Poverty

- **Relative poverty** is contextual. In the UK, households are below the poverty line if their income is 60% below median household income after housing costs for that year.

- **Absolute poverty** is measured against a set limit. People in absolute poverty are living below a certain income threshold that's considered necessary to meet basic needs, such as food, shelter, and clothing.

Employment

- Employment can be measured using the IMD dimension of employment deprivation, which measures the percentage of working-age people who would like to work but cannot (due to unemployment, sickness, disability, or care responsibilities).

- The census measures the percentages of those in full-time employment, in part-time employment, who are self-employed, and who are unemployed. (These are percentages of the economically active population – the people in work or seeking and available to work but who are without a job.)

Measuring social progress

- **Social progress** can be measured by reductions in inequalities, in improvements in social measures of deprivation, and in demographic changes. Income inequality in the UK rose rapidly in the 1980s. It has since broadly levelled off.

- Health is closely linked to living and working conditions and inequities (unfair conditions of access) in power, money, and resources. These are the social determinants of health. Successful regeneration can lead to improvements in life expectancy and reductions in health deprivation.

> **REVISION TIP**
>
> Remember that assessing success involves making complex judgements – schemes may be successful in some ways and not in others.

Regeneration and improvement in the living environment

Air pollution

- May be reduced by the introduction of clean air zones, which will help the UK to meet legally binding targets for emissions of gases such as nitrogen dioxide (NO_2).

- These can involve charges, such as London's Congestion Charge (2003), which has seen congestion reduced by 30%. The London ULEZ was introduced in 2019 and expanded to inner London in 2021. NO_2 levels have fallen by 46% in central London.

Abandoned and derelict land

Brownfield land, especially long-term abandoned or derelict land, is unattractive to developers, but governments have tried to incentivise its development.

- The 2016 Housing and Planning Act ordered councils to register brownfield land to help developers identify sites.

- The 2020 Town and Country Planning Order fast-tracked demolishing and reconstruction of buildings.

- The Scottish government's City Region Deal offered funding to develop derelict land, e.g. Raining's Stairs, Inverness.

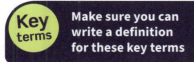 **Make sure you can write a definition for these key terms**

absolute poverty   relative poverty   social progress

⚙ KNOWLEDGE

5 Regenerating places

5.11 Judging the success of urban regeneration

Queen Elizabeth Park, Stratford, East London

A key factor in the London bid to host the 2012 Olympic Games was a vast urban regeneration project. This was overseen by the London Legacy Development Agency.

East London historically experienced higher levels of deprivation than west London (partly owing to prevailing westerly winds carrying air pollution from the city). Deindustrialisation meant that the industries, docks, and warehouses closed. The Lee Valley contained much abandoned and derelict land.

Sports-led:
UK and international tourists spending money that would circulate in the local economy.

Retail-led:
The Westfield Shopping Centre in Stratford would create thousands of jobs.

Strategies used in QEP urban regeneration

Property-led:
The athletes' village would become the East Village with 2818 homes, including 1379 affordable units.

Transport-led:
A new Stratford International station and new Elizabeth tube line for business and leisure travel would be built.

Contested nature of decisions on regeneration

These regeneration strategies were (and are) **contested** within local communities.

Concerns include:

The need for new houses	vs	Splitting up existing communities and a reluctance to let incomers change an area (**NIMBYism**)
The need for local economic revitalisation	vs	Increasing the ease of commuting to central London (skills drain)
The need for an improved retail offer	vs	The increase in prices due to higher-end shops and gentrification
The need to stem outflow of skilled labour	vs	The social and cultural challenges brought about by a rapid influx of newcomers

Key terms — Make sure you can write a definition for these key terms

contested   NIMBYism   stakeholder

Consequences of urban regeneration strategies

Was the London Olympic Park regeneration a success?

Economic success?

- Better connections, e.g. Stratford International is a ten-minute tube ride from Kings Cross.
- Around 20,000 jobs will be created by 2030.
- £9.3bn was invested in local infrastructure, including transport, energy, and telecoms, which makes operating businesses easier.
- The immediate cost of hosting the Olympics (£8.77bn) was outweighed by £14.2bn trade and investment in the UK economy between 2012 and 2014.

Social and demographic success?

- Around 100,000 jobs were created in construction and the operation of the Games, and tens of thousands of longer-term jobs.
- Around 33,000 homes are planned by 2026. However, only a quarter of those built by 2021 were 'affordable', and even these require salaries of greater than £60,000 a year, which makes them too costly for the original residents to rent or buy.
- The East Bank includes new outposts for the V&A Museum and UCL.
- Nearby Carpenters Estate is partly empty as redevelopment plans have stalled.
- West Ham FC now occupies the Olympic Stadium, bringing local pride to many.
- Some facilities offer long-term health benefits, e.g. the Aquatics Centre and velodrome.
- There is cycle- and pedestrian-friendly urban parkland.

Environmental success?

- There is 250 acres of new parkland, 8.35 km of new waterways leading to habitats, and cleaner air.
- Around 2.5 million cubic metres of contaminated soil were cleaned up on this brownfield site.
- Around 75% of the Games venues were repurposed, reducing post-event waste.
- Improved public transport leads to fewer CO_2 emissions and other private transport disbenefits.

Stakeholders assess success using different criteria

Stakeholder	Priorities
National government	Higher global status for the UK, tourism income, tax revenue
Mayor of London	Higher global status for London, improvements to East London, mass transit, and housing
Local council (Newham)	Housing, employment, skills, transport, sports amenities
Developers (McAlpine)	Profit, status, reputation
Local businesses	Trade, infrastructure, continuity
Local communities	Housing, employment, skills, transport, sports amenities

> **REVISION TIP**
>
> Remember, success may be judged using short- and long-term criteria.

Impact of change

The lived experience of those living, working, and playing in the regenerated site will depend on their personal context, including their daily encounters and their long-term social, cultural, and economic circumstances. The reality of life for residents may differ from the image cultivated by those who want to show off its successes.

Interest groups are heterogenous (varied). Local residents can be split into subcategories, including:

- Homeowners who took the opportunity to sell up when the Olympic buzz was high.
- Renters, whose rent increased due to increased demand.
- Younger residents who generally welcomed the economic, social, and cultural revitalisation.
- Older residents who tended to value local interactions and had developed deep roots.

⚙ KNOWLEDGE

5 Regenerating places

5.12 Judging the success of rural regeneration

North Antrim coast, Northern Ireland

The North Antrim coastline is seen as beautiful, windswept, and geologically fascinating. It is famous for the Giant's Causeway (Northern Ireland's only UNESCO World Heritage site), golf resorts, and film locations.

But it has experienced challenges, including remoteness, inaccessibility, the Northern Ireland 'Troubles,' and farm profitability.

It will face challenges of climate change – heatwaves, coastal erosion, etc.

▲ **Figure 1** *The Antrim coast starts in Belfast and ends at Londonderry*

◄ **Figure 2** *The Giant's Causeway in County Antrim, Northern Ireland, is a UNESCO World Heritage Site*

Sports-led: Portstewart golf course and other golf tourism attractions.

Travel-led: Promotion of the Causeway Coastal Route.

Heritage-led: Giant's Causeway, including £18m visitor centre (2013) and castles.

Strategies used in North Antrim rural regeneration

Diversification-led: Helping farms diversify into adventure-led tourism.

Food-led: Local food at Causeway market, Coleraine; branding of Menu of Moyle; whiskey.

Contested nature of decisions on regeneration

These regeneration strategies are contested within local communities. Concerns include:

The need for income from tourism	vs	The loss of local character as the hospitality industry increases investment
The need for local economic revitalisation	vs	Longstanding attachments to rural ways of life (e.g. cattle farming) regardless of their profitability
The demand for new golf resorts and the associated employment	vs	The loss of ecosystems (dunes and other coastal habitats)
The improved quality of the Causeway Coastal Route for drivers and coach companies	vs	Congestion and air/noise/visual pollution for all local residents, especially those with breathing difficulties (up to 85 coaches a day visit the Giant's Causeway)

Consequences of rural regeneration strategies

Have regeneration efforts on the North Antrim coast been a success?

Social and demographic success?

- Around 71% of working-age adults are economically active. This is lower than for Northern Ireland as a whole (73%). This shows that there is scope for more regeneration. Jobs in farming and manufacturing areas are falling.
- Educational attainment is in line with the rest of Northern Ireland.

Economic success?

- Tourism is seasonal and fickle. Travel restrictions (e.g. Covid-19), fashions, and economic trends impact visitor numbers.
- The failure of Bushmills Dunes golf resort reduces potential for job and income generation. But might the resort have put off some tourists who are drawn to the area's scenery?
- The borough contributes approximately £2bn annually to the Northern Ireland economy.
- Farms that have diversified and draw income from produce and day visitors are thriving.

Environmental success?

- The failure of Bushmills Dunes golf resort.
- The Giant's Causeway area is very well protected (UNESCO status, Area of Outstanding Natural Beauty, more than 90% of the land owned by the National Trust), but coastal paths are threatened by erosion.
- The Giant's Causeway visitor centre, designed to be unobtrusive in a nearby hillside.

Stakeholders assess success using different criteria

Stakeholder	Priorities
UNESCO	Maintain globally important geological and natural heritage
National government	Higher global status for the UK, tourism income, tax revenue
Devolved government (Northern Ireland)	Higher status, reduce rural inequalities, tax revenue, many tourists will spend time at other attractions in Northern Ireland (e.g. Titanic Museum, Belfast)
Local council (County Antrim)	Employment, skills, transport
Causeway Coast and Glens Borough Council	Encourage visitors to stay for more than a day trip, circulating more money in the local economy
Developers	Profit, status, reputation
Local businesses	Trade, infrastructure, continuity
Local communities	Housing, employment, skills, transport, sports amenities

Impact of change

The landscape has hardly altered; most notable changes are upgrades to roads, larger viewing areas, and more hotels. Lived experience will depend on personal context.

- Dairy farmers: some have diversified and others have stuck to their core business.
- Conservationists: have formed protest groups against new golf resorts.
- Younger residents: generally welcome the job opportunities presented by tourism and diversification.
- Older residents: tend to value local interactions.
- Small business owners and residents: pleased in peak tourist times, but experience quiet periods too.

REVISION TIP
Consider how wide the impacts of regeneration might spread.

RETRIEVAL

Learn the answers to the questions below, then cover the answers column with a piece of paper and write as many as you can. Check and repeat.

Questions / Answers

#	Questions	Answers
1	What three measures can be used to assess economic regeneration?	Income, poverty, and employment
2	What is relative poverty?	Poverty compared to others in the society
3	What is absolute poverty?	Living on an income beneath the threshold considered necessary to meet basic needs
4	What can reductions in inequalities, both between areas and within them, demonstrate?	Social progress
5	What health benefits can successful regeneration lead to?	Improvements in life expectancy and reductions in health deprivation
6	What improvements to the living environment can successful regeneration lead to?	Reduction in air pollution; a reduction in abandoned and derelict land
7	What name is given to previously developed land?	Brownfield
8	How is lived experience different from broader statistical success criteria?	It considers people's personal context, including their daily encounters and their long-term social, cultural, and economic circumstances
9	Why do proposals for new golf resorts often result in conflict?	Need for jobs and investment versus loss of dune habitat

Put paper here

Previous questions

Now go back and use these questions to check your knowledge of previous topics.

Questions / Answers

#	Questions	Answers
1	Name the three broad types of international immigration policy.	Open doors, closed doors, targeted
2	Why might gated communities be problematic?	Social uniformity (within estate), and deepening social segregation and suspicions of others
3	What are two benefits of Enterprise Zones for businesses?	Two from: business rate discounts / tax relief on machinery / simplified local planning
4	What is a partnership-led regeneration strategy?	One that relies on funding from public and private sources
5	Give two ways in which deindustrialised urban areas can use their heritage to rebrand.	Two from: museums / visitor centres / awesome structures / use of derelict land to host buildings or events

Put paper here

PRACTICE

Exam-style questions

1 Suggest **one** reason why regeneration strategies often involve tensions. **(3)**

2 Suggest **one** reason why local government efforts to attract businesses might have varying levels of success. **(3)**

3 Study **Figure 1**. Suggest **one** reason why conflicts can occur among contrasting groups in communities about the priorities for regeneration. **(3)**

> **EXAM TIP**
>
> Before you start writing your answer, make sure that you BUG the question:
> - Box the command word.
> - Underline the key terms.
> - Glance at the question again.

> Kenyon Park* has been used by groups of young people, some of whom are drinking alcohol and using drugs.
>
> Unfortunately, incidents of anti-social behaviour from such groups are growing again. Loud music has been played late at night, and litter associated with drinking and drug-taking has been found in the more isolated places in the park, especially in the bottom scrub area and on the wooded slope.

*Name and other details have been changed

▲ **Figure 1** *Excerpt from website of Friends of Kenyon Park* (in the suburbs of a large UK city)*

4 Suggest **one** reason why local interest groups seek to influence decision-making about regeneration projects. **(3)**

5 Study **Figure 2**. Suggest **one** reason why places have changed their demographic characteristics over time. **(3)**

> **EXAM TIP**
>
> Three-mark questions require you to give one reason, then make two points of development. Think of these as 'reason, so, so' answers.

Region	Black British, Asian British, and mixed	White British and White other
	%	%
East Midlands	13	86
East of England	12	87
London	41	54
North East	6	94
North West	12	85
South East	12	87
South West	6	93
Wales	6	94
West Midlands	21	77
Yorkshire and The Humber	13	86

▲ **Figure 2** *Areas of England and Wales by ethnicity, 2021*

Exam-style questions

6   Explain why local governments compete to attract domestic and foreign investors. **(6)**

7   Explain why places change their demographic characteristics over time. **(6)**

8   Explain why inequalities in pay levels and types of employment may result in differences in quality of life between places. **(6)**

9   Explain why some regeneration strategies are more contested than others. **(6)**

10   Study **Figure 3**. Suggest how the rate and type of development (such as house building targets and housing affordability) might influence economic regeneration. **(6)**

> **EXAM TIP** 🎯
>
> Some questions specify urban or rural regeneration, while others are open-ended. Ensure you read the question carefully.

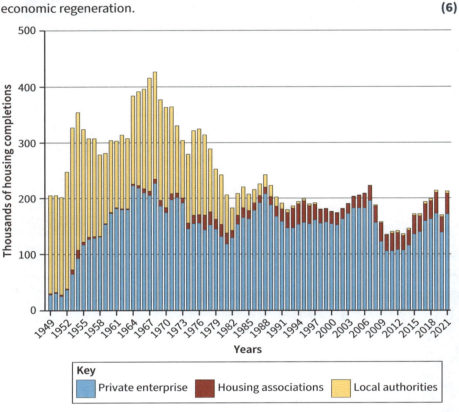

◀ **Figure 3** *New homes completed by private companies, housing associations, and local authorities in the United Kingdom from 1949 to 2020*

11   Study **Figures 4a** and **4b**. Explain why changes to the living environment can influence the success of regeneration. **(6)**

▲ **Figure 4a** *Bedford Street, Middlesbrough, before regeneration (2016)*

▲ **Figure 4b** *Bedford Street, Middlesbrough, after regeneration (2017)*

12 Study **Figure 5**. Explain why some UK deindustrialised cities use rebranding as a regeneration strategy. **(6)**

> Leeds has embraced the concept of culture-led regeneration. This refers to attempts to use visual and performing arts, music, and literature as ways to attract people and investment to an area, whilst fostering a sense of local pride. As well as the Royal Armouries, it hosts a cultural quarter which is home to the Leeds Playhouse, BBC Yorkshire, Northern Ballet, and the Leeds College of Music. Leeds was midway through the bidding process to become an EU Capital of Culture when the UK left the EU, so the council decided to proceed with its own 'year of culture' – Leeds 2023.

▲ **Figure 5** *Information about culture-led regeneration in Leeds*

13 Explain how lived experience can influence perceptions of the success of regeneration strategies. **(6)**

14 Evaluate the impact of government decisions on the economic and social characteristics of either urban or rural areas. **(20)**

15 Evaluate the view that different urban stakeholders have different criteria for judging the success of urban regeneration. **(20)**

16 Evaluate the range of ways which can be used to determine whether your local place needs regeneration. **(20)**

EXAM TIP

Consider ending responses to 'Evaluate' questions with a view about how things may change in the future.

6 Diverse places

6.1 Changing populations

The population of the UK

The 2021 census showed that the UK population had reached 67 million, continuing the upward trend seen since the Second World War (with some stagnation in the 1970s). Recent years have seen a continuation of the trends seen since the 1970s.

OS maps can be used to show how populations change over time, as they show provision of schools, hospitals, places of worship, place names, named museums, and other cultural places.

REVISION TIP

Use the up-to-date, and often interactive, population data provided by the ONS to help you revise.

Some areas have a shrinking population, including north-west Wales (less labour-intensive pastoral farming and fewer economic opportunities).

The population is growing more slowly in parts of rural northern England where manufacturing (e.g. shipbuilding) or mining (e.g. coal) were traditionally more prevalent.

The populations of some large cities (e.g. Leeds and Manchester) fell in the 1970s and 1980s, but have since risen due to:

- urban **regeneration** programmes
- an increase in the number of university students
- rebranding of city centres as being fashionable
- investment by property developers in apartment blocks.

The population has grown rapidly in some areas (especially London and the south-east).

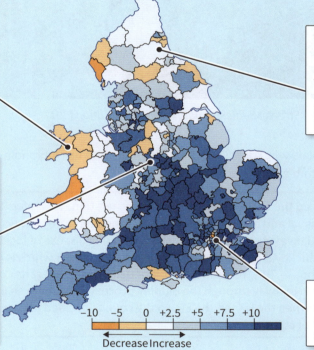

−10 −5 0 +2.5 +5 +7.5 +10

Decrease Increase

▲ **Figure 1** *Population change in England and Wales from 2011 to 2021*

Population structure and density

- The average (median) age in the UK is 40.7 years.
- The oldest average age (44.1 years) is in south-west England, while the youngest average age (35.9 years) is in London.

- Populations tend to be younger in larger, urban areas, and older in more rural areas.
- **Population density** is highest in inner-city areas and lowest in remote areas.

Accessibility: how easy it is for residents to travel to and from residential, commercial, and other zones.

Historical development: developments will remain for several decades or centuries before demolition.

Why does population structure and density vary?

Physical factors: barriers (hills, wetlands, rivers) and climate (open or elevated areas exposed to extremes).

Planning: historical and present-day laws constrain housing location, density, affordability, green space, etc.

How do places vary?

Inner city
Highest density: a result of industries and terraced housing built within walking distance of factories during the Industrial Revolution era, post-war immigration, and planning restraints of the Green Belt (outside built-up areas) which forced developers and councils to build in inner cities.

CBD (central business district)
Few residents: the area is mainly commercial, retail, and hospitality. There are some city centre residents, mostly in flats that sometimes date from 1980s renewal projects (e.g. led by urban development corporations).

Rural–urban fringe
Low density: mostly around old market towns and train stations, then later growth due to car ownership. Building in this area is limited by Green Belt policies. Historically home to commuters (dormitory settlements).

Suburbs
Medium density: a mixture of private and local authority housing, often initially spreading from hubs of mass transit systems (e.g. underground, trams, suburban trains, and bus routes) with later expansion and infilling from car users. Private developers are attracted by high profits from high demand, often from families. Suburbs also contain villages and towns incorporated by urban sprawl.

More remote rural areas
Some popular larger towns, still growing due to families, affluent residents, and retirees looking for a more pleasant living environment (clean air and green space). Some declining villages.

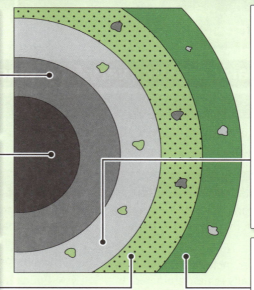

▲ *Figure 2* **Population structure** *and density vary according to placement in the* **rural–urban continuum**

REVISION TIP
Keep in mind different rates of change *within* regions as well as *between* regions.

Population structure and dynamics

People of working age from the UK (**internal migration**) and other countries (**international migration**) seek employment in cities, leading to a bulge in population pyramids around the 20- to 40-year-old age bracket, especially for inner city and diverse communities.
Fewer older immigrants and premature mortality (due to poor housing, air quality, diet, etc.) lead to narrow peaks in population pyramids.

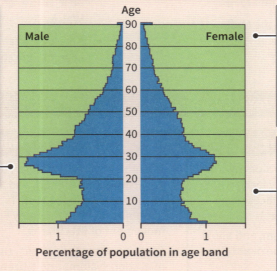

▶ *Figure 3* *A population pyramid for an inner London borough*

Mortality rate in the UK is fairly low (10 per 1,000 people per year); life expectancy is high. In the 2010s, life expectancy in England stalled, and the amount of time spent in poor health increased. Similar trends occurred in Scotland.

Fertility rates in economically developed countries like the UK are low (average 1.6 children per woman). The base of the population pyramid is narrow. Fertility rates are higher in communities with a high proportion of migrants from developing countries, but these tend to fall over time.

UK facts

- Life expectancy is 8.6 years lower in the most deprived 10% of communities than in the the least deprived.

- Fertility rates in rural areas tend to be lower, as there are fewer people of child-rearing age, but life expectancy tends to be higher (less pollution, and sometimes more affluent, depending on the area).

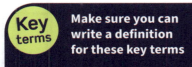

Key terms
Make sure you can write a definition for these key terms

fertility rate   internal migration   international migration
mortality rate   population density   population structure
regeneration   rural–urban continuum

⚙ KNOWLEDGE

6 Diverse places

6.2 Population characteristics vary from place to place and over time

Gender

In UK settlements, the number of men and women is usually even.

However, there are some exceptions:

Rutland, England

In 2021, Rutland unitary authority had more men than women: for every 100 women, there were 106 men.

Reasons for disparity:

- Agriculture is a big employer, including farm labour, and this industry is historically male-dominated. It now attracts single male migrants working in food growing and processing.
- Many young women move away to go to university, and many do not return.

Camden, London

In 2021, the London borough of Camden had more women than men: for every 100 men there were 104 women.

Reasons for disparity:

- Educational institutions (e.g. University of the Arts London) have a large female student population.
- Camden's reputation as an inclusive, open-minded, and safe area may attract female residents.

Ethnicity

Ethnicity can vary hugely between settlements.

- Cities tend to be more ethnically diverse than smaller settlements and rural areas.
- London, and some cities in the Midlands and north of England, are very diverse.

- Other places are less diverse, e.g. most of the north-east, south, and south-west of England, and most of Wales, Scotland, and Northern Ireland.
- This is a dynamic situation. Some settlements have seen rapidly changing ethnicities due to immigration.

Why does cultural diversity vary?

Social clustering

- **Social clustering** usually occurs by choice, to share community facilities (e.g. shops, food outlets, places of worship, community centres)
- Also to avoid possible racism, and allow language and customs to be maintained.

Planning policy

- After the Second World War, the UK government sponsored job advertisements abroad, and agreed immigration targets, to fill labour gaps in the NHS, factories, and public transport. EU membership brought many Europeans to the UK through free movement of EU citizens.
- Some government actions, such as closed-door or points-based migration systems and citizenship tests, suppress diversity.

Physical factors

- These play little part in **cultural diversity**, although some people who are in the minority ethnically report being made to feel 'out of place'.

Accessibility to key cities

- Cities had higher employment opportunities when earlier migration occurred.
- Train links helped migrants to move, and also to connect with members of the same diaspora in other cities.

Key terms Make sure you can write a definition for these key terms

cultural diversity   ethnicity   gender   social clustering

Bradford, England	Newham, London
In 2021, Bradford's population identified as:	In 2021, Newham's population identified as:

Bradford, England

In 2021, Bradford's population identified as:

- 32.1% Asian, Asian British, or Asian Welsh (26.8% in 2011)
- 25.5% Pakistani
- 61.1% White (67.4% in 2011)
- Significant number of Eastern Europeans.

Reasons for diversity:

- From the 1950s to 1970s, there was a wave of immigration, particularly from Pakistan and India, to work in the textile and manufacturing industries.
- When Mangla Dam in Mirpur, Pakistan, was built in 1965, thousands of people were displaced. The compensation they received enabled them to travel to Bradford to take up housing and jobs in the textile industry, leading to chain migration.
- Close-knit families and communities have retained their cultural and ethnic identities.

Newham, London

In 2021, Newham's population identified as:

- 42.2% Asian, Asian British, or Asian Welsh (down from 43.5% in 2011)
- 30.8% White (up from 29% in 2011)
- 17.5% Black, Black British, Black Welsh, Caribbean, or African (down from 19.6% in 2011).

Reasons for diversity:

- Proximity to London's financial and economic hub has attracted employment and educational opportunities.
- The Olympic legacy and University of East London offer employment and training.
- History of being a welcoming area for immigrants and refugees.
- The appeal of cultural vibrancy and diverse array of businesses and cuisines.

Fertility and mortality rates

Fertility and mortality rates shape the cultural characteristics of places.

Decline in fertility rates leads to smaller family sizes, impacting local community dynamics.

In rural areas and most urban areas, lower fertility rates mean difficulties in maintaining social infrastructure (e.g. schools and healthcare facilities).

In some urban places, higher fertility rates result in higher demand for housing and services.

Impacts of differing fertility and mortality rates

Longer life expectancies leads to an aging demographic. This can slow down cultural change and dynamic reaction to culture shifts.

An ageing population may encourage more preservation of traditions, and increased demand for healthcare and elderly care services.

International and internal migration

International and internal migration contribute to cultural diversity.

International migrants (to mainly larger cities) bring traditions and practices to new places, leading to a mixing of ideas.

Internal migrants from urban to rural areas seeking a quieter lifestyle can spread 'urban' norms.

In both urban and rural areas, these migration patterns can lead to the formation of ethnic enclaves or cultural fusion, influencing local traditions, cuisine, languages spoken, and overall cultural identity.

Comparing change over time

It can be difficult to compare changes over time owing to survey design and differences in the way individuals chose to self-identify between censuses. For example, more people identified as 'English' in the 2021 census than in 2011, but this may have been because the 'English' option was placed above the 'British' one.

REVISION TIP

Be careful not to overgeneralise when referring to groups.

⚙ KNOWLEDGE

6 Diverse places

6.3 Past and present connections shape demographics

Influences on places

Global influences

- Has the global shift of manufacturing and certain services abroad, especially to Asia, affected your places?
- Have global population movements, conflicts, or political, cultural, or environmental trends affected your places?

International influences

- What **TNCs** have headquarters or other operations in your places?
- Have your places received funding or investment from the EU or other **IGOs**?
- Are there any destinations which attract international tourists and/or students in your places?

National influences

- What national businesses have headquarters or other operations in your places?
- What national government initiatives have taken place in your places?

Regional influences

- What local businesses have their headquarters in your places?
- What local government initiatives have taken place in your places?

> **REVISION TIP**
>
> Remember that your chosen and contrasting places should be a locality, a neighbourhood, or a small community, *not* an entire city.

> **REVISION TIP**
>
> Summarise the characteristics of your chosen places to help you remember them. Consider demographics (e.g. percentage under 16, over 60, British, ethnic identities, languages spoken), economics (e.g. employment structure), culture, land use, and the built and natural environment.

Places can be represented in different ways

- How has the population changed in terms of age, gender, and ethnicity?
- How have ways of life changed? Think about religion, language, food, drink, dress, customs, etc.
- What have people lost (**cultural erosion**) or gained (**cultural enrichment**)?
- How do people feel about the changes?

How the lives of students and others are affected

- How are the lives of students affected by continuity and change in your places?
  — Where do students spend any available free time?
  — Where do students work?
  — How do students get around?
- How are the lives of other people affected by continuity and change in your places?
  — Where do other people spend any available free time?
  — Where do other people work?
  — How do other people get around?

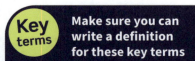 **Make sure you can write a definition for these key terms**

cultural enrichment   cultural erosion   IGO   TNC

RETRIEVAL

Learn the answers to the questions below, then cover the answers with a piece of paper and write as many as you can. Check and repeat.

	Questions	Answers
1	Name one way that OS maps show how populations change over time.	One from: provision of: schools / hospitals / places of worship / place names / named museums / other cultural places
2	What is happening to the size of the population of north-west Wales?	Shrinking
3	The populations of some large cities declined in the 1970s and 1980s. Give one reason why their populations have since risen.	One from: urban regeneration programmes / increase in number of university students / rebranding of city centres as being fashionable / investment by property developers in apartment blocks
4	Which UK region has the youngest average age?	London (35.9 years)
5	Name the four key reasons why population structure and density vary.	Accessibility, physical factors, historical development, planning
6	Rank these areas from highest to lowest residential population density: suburbs, rural–urban fringe, inner city.	Inner city, suburbs, rural–urban fringe
7	Are fertility rates in economically developed countries such as the UK high or low?	Low
8	How do fertility rates in areas recently occupied by international migrants compare with those of the rest of the UK?	They tend to be higher than the UK average (over time, they get closer to the UK average)
9	Give the three key reasons why cultural diversity varies.	Social clustering, accessibility to key cities, planning policy
10	Why is Bradford ethnically diverse?	Jobs in textile industry, chain migration (due to being able to pay to travel to the UK using compensation funds for dam in Pakistan)
11	Name four aspects of culture which might be affected by change in an area.	Four from: religion / language / food / drink / dress / customs
12	Would the loss of a community centre be cultural erosion or cultural enrichment?	Cultural erosion

Put paper here

⚙ KNOWLEDGE

6 Diverse places

6.4 Perception of urban places

Perceptions of urban places during industrialisation

In the nineteenth century, urban factories attracted workers, but the industrial zones of UK cities were perceived as overcrowded, polluted, dangerous, and threatening.

In the 1960s and 1970s, cities were perceived as dynamic, positive areas, rising out of the ruins of the Second World War bombing raids and 1950s austerity. Nevertheless, falling housing densities and outmigration meant continued shrinking.

In the 1980s, some city centres and inner-city areas were called ghost towns due to deindustrialisation and depopulation.

Since the early 2000s, there has been a sharp rise in population, because people have been attracted to Manchester's economic opportunities and variety of social and leisure activities resulting from regeneration programmes.

◄ **Figure 1** *Since 1800, the population of Manchester has risen and fallen as perceptions of the city have changed.*

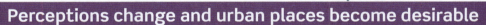

Perceptions change and urban places become desirable

Since the 1990s, many cities have been seen as attractive to certain groups, especially young people and migrants, despite ongoing challenges relating to inequality.

Economic centres: there has been a shift from the old economy to the new economy (services, high-tech, finance, creative centres).

Cultural diversity: due to constant migrant flows.

Education: universities and colleges bring an influx of national and international students and foster innovation

Attractions of urban places

Infrastructure: mass transit, fast broadband, etc. give a high-tech and efficient image.

Leisure: nightlife and café culture offer opportunities for socialising and networking.

Innovation and creativity: there is collaboration and cross-pollination of ideas among diverse groups.

Culture: museums, theatres, art galleries, festivals.

Perceptions of urban locations

Some urban locations are perceived as undesirable. Reasons include perceptions of:

- Crime rates, especially street-level crime.
- Environmental quality (e.g. litter, broken windows).

- Reputation as a 'no-go area', particularly at night, can be hard to shift.
- Population characteristics. For example:
  — Young people in groups can be perceived as undesirable.
  — Ethnic diversity and social clusters may lead to some feeling that a place has changed too much.
  — Inequality in income cause some to feel lack of commonality.

REVISION TIP

Remember that stereotyping can lead to prejudice and discriminatory behaviour.

Reputation based on quantitative data and lived experience

Quantitative data	Lived experience
For example, from the UK police website (recorded crime), census data, surveys, IMD data (income, employment, health, education, crime, and living environment).	Everyone's **lived experience** of place is unique.

Age: young people may see or experience more street-level crime (e.g. drug use) than retirees.

Ethnicity: members of different groups may experience discrimination.

Length of residence: long-standing residents will know more about an area.

Aspects of urban lived experience

Gender: cis gender men may feel safer than people of other genders.

Levels of deprivation: affluent residents or outsiders may spend less time 'on the street' and may form more of their opinions from media. Less affluent residents, may have more first-hand experience.

Sexuality: members of the LGBTQIA + community may perceive some spaces as safer than others.

REVISION TIP

Remember these factors overlap and are dynamic (change over time). Societal trends and attitudes also change. To delve deeper into place attachment and engagement, refer to your own experiences or those of people you have talked to.

Reputation based on media representation

- Newspapers aim to gain and retain readership by highlighting interesting stories about places. These may be unrepresentative of everyday life in the place, but such impressions can last.
- Sensationalised articles labelling a place as the 'worst place to live in the UK or referring to no-go' areas act as clickbait in online media.
- Different media will tend to show places in different lights: national newspapers, local newspapers, and social media will contribute to different perceptions of a place.
- Some media may perpetuate stereotypes of urban areas.

REVISION TIP

Use visual data to examine representations. For example, you could make an annotated collage of media representations.

Perceptions of suburban and inner-city areas

Groups	Suburbs	Inner-city areas
Age	Parents may like them due to proximity, green space, and perceived safety. Younger adults may feel they lack excitement and jobs.	Younger people may like the walkability, 'buzz', social opportunities, and affordable housing.
Ethnicity	Some suburbs are not diverse. But others, especially inner suburbs, have become more culturally diverse.	Some may be social clusters. Others are more culturally diverse.
Life-cycle stage	Some suburbs offer work opportunities in the service sector, more specialised jobs may be in the CBD. Working from home may change perceptions.	Younger workers can easily commute to jobs nearby or in the CBD. Retirees may want more space and quiet.

REVISION TIP

Be careful not to exaggerate or sensationalise the challenges, and perceptions of such challenges, that some areas face.

 Key terms Make sure you can write a definition for these key terms

industrialisation   life-cycle stage
lived experience
media representation
quantitative data

6 Diverse places

6.5 Perception of rural places

Perceptions of rural places as idyllic

Tranquil environments: lower population density means less noise and pollution.

Natural landscapes: most rural areas in the UK (e.g. Lake District) have been cleared of forests, and are intensively managed by humans.

Historical significance: preserved architecture, ancient ruins, and heritage sites inspire nostalgia and romanticism.

Perceptions of rural areas as idyllic

Cultural associations: traditional ways of life foster a sense of cultural authenticity.

Sense of community: smaller populations and tight-knit communities (social cohesion).

Connection to nature: through outdoor activities like hiking, fishing, and stargazing.

Cleaner air and environment: few manufacturing industries so a healthier and more appealing environment.

Hardy's Wessex

The ancient kingdom of Wessex inspired author Thomas Hardy to romanticise rural areas, emphasising their natural beauty and cultural significance in his novels. He highlighted rural tranquillity, suggesting a harmonious connection between the natural world and human experience.

REVISION TIP

Look at photos of different types of rural areas, and apply the factors listed here to consider how groups may form their perceptions of such areas.

Perceptions of some rural locations as undesirable

Negative perceptions of some rural areas include:

- Remoteness and isolation: difficulties in accessing essential services, social activities, and job opportunities.

- Limited social opportunities: fewer social activities, cultural events, and entertainment options.

- Limited range of services: fewer healthcare facilities, educational institutions, and shopping centres.

- High transport costs: due to longer distances and limited public transportation.

- Population characteristics: limited youth and job prospects, meaning less vibrancy and diversity.

- Infrastructure challenges: narrow roads, unreliable internet, and limited utilities (e.g. often no mains gas).

REVISION TIP

Remember that rural and urban residents will often differ in their perceptions of rural places.

Reputation based on quantitative data and lived experience

Quantitative data can be accessed as for urban areas (see 6.4).

Age: older children may perceive a lack of social interactions and activities.

Length of residence: long-standing residents will know more about an area.

Levels of deprivation: lower-income residents find transport costs challenging and public transport inadequate.

Aspects of rural lived experience

Ethnicity: some people report feeling out of place in less culturally diverse rural areas.

Gender: traditional gender roles may be more common in remoter rural areas, which may not be in line with the views of new arrivals or outsiders.

Sexuality: members of the LGBTQIA+ community may perceive there to be fewer dedicated or openly welcoming spaces.

Reputation based on media representation

- National newspapers often romanticise rural areas, ignoring the challenges and making some residents feel unseen.
- Social media may encourage tourism (e.g. to Instagrammable locations), which can simplify and exoticise rural experiences.
- Photography often shows quiet, 'unspoilt' areas, omitting rural residents and their lived experiences.
- Some media can perpetuate negative stereotypes of areas.

> **REVISION TIP**
>
> Try not to assume too much based on third-party sources. If possible, visit rural areas and collect and analyse your own data for practice.

Rural areas are viewed in different ways

Attitudes may vary within groups of residents or visitors, according to age, employment, life stage, level of education, etc.

Commuter villages and retirement villages

Commuter villages and **retirement villages** are often found in the rural–urban fringe.

Advantages	Disadvantages
☑ Accessible to urban and rural areas. ☑ Commuters often appreciate a physical gap between work and home. ☑ Retirees appreciate access to green spaces and ease of access to retail and community spaces. ☑ Open space allows development of retirement villages (bungalows) or complexes (apartments, with gardens and other services on site).	☒ High house prices. ☒ High traffic volumes (especially where there are limited public transport options).

Remote rural areas

- These may escape some of the challenges of rural-urban fringe areas. For example, they have less traffic and lower house prices.
- They are often seen romantically by outsiders. This can lead to conflict: residents who want to develop their homes, farms, or businesses may clash with some visitors who expect places to be preserved rather than conserved.

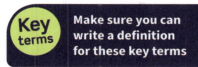 **Make sure you can write a definition for these key terms**

commuter village   retirement village

6 Diverse places

6.6 Perceptions of your local place

Data for your chosen places

Category	Statistical evidence to consider
Economic	Average house prices
Social	Police.uk website: anti-social behaviour and other incidents Census: • % of residents with no educational qualifications • self-reported levels of health IMD decile for: • deprivation of education, skills, and training • health deprivation and disability • barriers to housing and services • crime
Environmental	IMD decile for living environment deprivation Green space: magic.defra.gov.uk Air pollution: naei.beis.gov.uk (emissions map)
Composite	IMD: What rank is your LSOA on the list of 32,844 LSOAs? What decile (10% band) does this place your area in?

REVISION TIP

Research and analyse data for both your chosen places to determine whether people have positive of negative perceptions of them. Consider the points listed on this page.

REVISION TIP

Read news stories relating to your local place. Look at a broad range of media, not just one or two outlets. Search for news rather than let it come to you through your social media filters.

Media presentations of places

Different media can provide contrasting evidence about perceptions of your chosen places.

National newspapers

- Often sensationalise
- Have political angles
- May be populist
- May be more elitist or economically minded
- May be more mindful of social justice issues
- Often show lists of 'best' and 'worst' places to live or visit in Britain, based on selective evidence

Local newspapers

- May sensationalise
- May provide grounded and balanced representations of places, covering good and bad news
- Report community events, showing levels of engagement
- Often celebrate local areas
- Publish letters about local issues

Social media

- May sensationalise
- Can show polarised views at expense of middle ground

Other media

- Photography can highlight the human angle. It often shows unusual events
- Local TV news tends to celebrate places more than national news
- Documentaries are in-depth but they may exaggerate places
- Film, music, literature, art – formal and informal (e.g. graffiti)

⇄ RETRIEVAL

Learn the answers to the questions below, then cover the answers with a piece of paper and write as many as you can. Check and repeat.

Questions

Answers

	Questions	Answers
1	Give the two main reasons why some inner-city areas were known as ghost towns in the 1980s.	Deindustrialisation and depopulation
2	Give two reasons why some urban locations are perceived as undesirable.	Two from: crime rates / environmental quality / reputation / population characteristics
3	Name two sources of quantitative data that can affect perceptions of urban areas.	Two from: recorded crime / census data / surveys / Index of Multiple Deprivation (IMD)
4	What term means the ways that people encounter places on a day-to-day basis?	Lived experience
5	Name two factors that can affect lived experience.	Two from: age / length of residence / levels of deprivation / ethnicity / gender / sexuality
6	Which fictional county did Thomas Hardy write about, romanticising its rural areas?	Wessex
7	Name two infrastructure challenges experienced by many rural areas.	Two from: narrow roads / unreliable internet / limited utilities
8	How might Instagram influence perceptions of rural areas?	It may simplify and exoticise rural experiences
9	Where are commuter villages often found?	Rural–urban fringe
10	Name three attractions of urban areas.	Three from: economic centre / cultural diversity / education / culture / infrastructure / leisure / innovation and creativity

Put paper here

Previous questions

Now go back and use these questions to check your knowledge of previous topics.

Questions

Answers

	Questions	Answers
1	Name the four key reasons why population structure and density vary.	Accessibility, physical factors, historical development, planning
2	Rank these areas from highest to lowest residential population density: suburbs, rural–urban fringe, inner city.	Inner city, suburbs, rural-urban fringe
3	Are fertility rates in economically developed countries such as the UK high or low?	Low
4	Give the three key reasons why cultural diversity varies.	Social clustering, accessibility to key cities, planning policy
5	Name one way that OS maps show how populations change over time.	One from: provision of: schools / hospitals / places of worship / place names / named museums / other cultural places

Put paper here

6 Diverse places

6.7 UK migration flows

Internal migration patterns

Highly educated young people are more likely than any other demographic to move within the UK because:

- skilled graduate jobs tend to be clustered in cities
- cities offer social and cultural attractions
- they often have experience of moving (to university).

Many are attracted to London and the south-east by the graduate jobs, high pay, and strong social and cultural opportunities.

Many adults in their 30s and 40s (often with young children) move away from London or other cities (**counterurbanisation**), usually to accessible rural areas, due to:

- the high cost of renting or buying property in the capital
- a desire for improved environmental quality (cleaner air, green spaces)
- a desire to avoid proximity to deprivation, and perceptions of crime
- the need for improved social and educational opportunities for their children.

Other key patterns in the 2010s and 2020s

- Growth of large northern cities (e.g. Manchester, Liverpool, and Leeds) linked to regeneration.

- Continued rural-to-urban movement (**urbanisation**) for 17 to 25 year olds (for education, jobs, etc.).

- Continued net migration from Scotland to England.

- Movement of retirees to rural areas, especially coastal, and semi-retired, helped by working from home and technology.

- Student migration as numbers of students grew in the 2010s, moving to large cities and smaller settlements with large universities (e.g. Oxford and Aberystwyth).

Uneven demographic and cultural patterns

Student-dominated area: Manchester
- Higher-education institutions, including Manchester and Salford universities.
- In 2022, one in eight of the population were students.
- Vibrant cultural scene with art galleries, cultural events, and large entertainment venues.

Older population, coastal town: Eastbourne
- Many retirees in owner-occupied homes, retirement communities, and care homes.
- In 2021, the median age in Eastbourne was 45 (national average is 40).
- The focus is on leisure activities such as golf, bowls, and traditional seaside entertainment.

◀ **Figure 1** Demographic and cultural patterns in England

Young urban population: Shoreditch, London
- Known for its younger and diverse population.
- In 2021, the median age in Hackney (which includes Shoreditch) was 32.
- A hub for creative industries, startups, and technology companies, street art, bars and clubs, and cultural events.

Manchester

Shoreditch

Eastbourne

International migration flows into the UK

The UK government has been the main **gatekeeper** affecting international migrant flows.

Pro-migration policies

1940s to 1960s: UK government promoted large-scale immigration from countries that had been colonised by Britain and sponsored migrants to fill certain job vacancies (e.g. in NHS).

Chain migration

1950s to present: once established in the UK, a migrant could apply for family members to join them.

By 2020, non-EU born people made up 9% of the UK total.

Advantages

- ☑ Migrants enrich the UK's cultural diversity and pluralism (cuisines, languages, traditions, festivals).
- ☑ There is enhanced social integration (cultural exchange, interfaith dialogue, fusion of traditions, unique British multicultural identity).
- ☑ Growth of UK's labour force (employees, entrepreneurs, trade and investment ties).

Free movement

1992: free movement for workers within the EU.

2004: eight Eastern European countries joined the EU (the largest being Poland).

2007: Bulgaria and Romania joined the EU.

EU arrivals

Many workers, especially from areas of Europe with lower pay, arrived in the UK.

By 2020, EU-born people made up approximately 5% of the UK population.

Challenges

- ☒ Integration policies, discrimination, resource demand, divisions exploited by some media.

Brexit

2016: following the Brexit referendum, some Eastern Europeans left the UK, and some stayed.

2021 saw the end of the UK/EU free movement and the introduction of a points-based system of eligibility.

EU departures

By 2022, the number of non-EU overseas workers (2.7 million) surpassed that of their EU counterparts (2.5 million).

Consequences

The character of some neighbourhoods is changing as EU populations slowly fall and non-EU populations from a wider variety of places (beyond South Asia and the Caribbean) rise.

Social challenges and opportunities

Some international migrants, especially East Europeans, have moved to rural areas, largely since 2004, due to employment opportunities in the agriculture, manufacturing, and food-processing sectors.

> **REVISION TIP**
>
> The word 'issues' is often used informally to refer to problems or challenges. It is a neutral word when used in examinations, so ensure you know enough to write a balanced account.

Opportunities

- ☑ They stimulate the local rural economy by increasing crop yields and manufacturing productivity.
- ☑ They address labour shortages during peak seasons, helping farmers and employers thrive. This increases taxes and, consequently, social investment.
- ☑ If the migrant community settles, this can help rural schools and services thrive.
- ☑ Sharing knowledge can diversify culture.

Challenges

- ☒ Integration can be difficult due to e.g. language and cultural differences.
- ☒ Can cause a demographic imbalance, as international migrants are often young and male.
- ☒ Some activities of international migrants may be perceived as being antisocial (e.g. late-night socialising).

East Europeans in Lincolnshire

Extensive agricultural activity has attracted migrants. This has boosted the local economy, but also presented challenges relating to language and cultural integration, highlighting the need for community building initiatives.

Key terms Make sure you can write a definition for these key terms

counterurbanisation   gatekeeper   urbanisation

⚙ KNOWLEDGE

6 Diverse places

6.8 Impacts of migration

Segregation

International migrants often choose to settle in distinctive places, with segregation closely related to economic and social characteristics.

Economic indicators	Social indicators
• Income and employment opportunities play a significant role in determining where international migrants settle.	• Social indicators such as health, crime rates, and educational opportunities also influence settlement patterns.
• Many Afro-Caribbean migrants who were recruited to drive London buses in the 1950s chose to live near bus depots, such as Brixton.	• Migrants may seek areas with better access to healthcare, lower crime rates, and quality education for their families.
• Many migrants from Pakistan to Bradford in the 1960s settled in areas within walking distance of the textile mills where they worked.	

Diverse living spaces in urban areas

Urban areas often exhibit diverse living spaces characterised by the presence of multiple ethnic and cultural groups.

These areas reflect the city's multiculturalism and may have distinct social features, including:

- a wide range of retail outlets, international grocery stores, and specialty shops and services
- places of worship serving specific religious communities
- leisure and cultural centres (e.g. community halls).

Southall, West London

Southall has a significant South Asian community, mostly of Indian and Pakistani descent.

- Vibrant retail scene, featuring numerous Indian and Pakistani grocery stores, clothing boutiques, jewellers, and spice markets.
- Places of worship: gurdwaras (Sikh temples), Hindu temples, and mosques. These are hubs of worship, community outreach, charity, and education.
- Cultural events like Diwali celebrations, Vaisakhi festivals, and Eid al-Fitr gatherings, reflecting the cultural diversity of the area.

Southall demonstrates successful social integration, with different ethnic groups co-existing and collaborating on community initiatives. This reflects the largely harmonious blending of different cultures.

> **REVISION TIP** 📋
> Try to summarise smaller case studies on a single sticky note each.

Experiences and perceptions of living spaces

- Experiences and perceptions of living spaces in the UK evolve as communities progress economically and culturally over generations. Younger generations often have different attitudes and norms to their predecessors.

- Global cultural trends, influenced by factors such as technology, migration, and globalisation, impact **intergenerational** attitudes, and lead to shifts in how people perceive and interact with their living spaces.

- The diversity brought about by globalisation (including multicultural neighbourhoods and blending of cultural influences) tends to be more celebrated by younger generations, especially those with mid and high incomes.

- Younger generations are often more environmentally conscious, so seek living spaces designed with sustainability in mind, influencing urban planning and housing development.

REVISION TIP

You know that not all young people will have the same experiences and perceptions of places. So, keep this in mind when considering their views!

2020 National Housing Federation survey	2019 VisitBritain survey
• Younger generations in the UK, particularly millennials and Generation Z, prioritise urban living and proximity to amenities over the larger living spaces sought out by their parents' generation. • Younger generations often prefer renting to buying homes (or are constrained into doing so).	Younger UK residents are more likely to engage more often in cultural activities (such as visiting museums, galleries, and live music events) than older generations.

Historical transformation:
Known for its industrial character and working-class population, it has gentrified since the 1980s, attracting young professionals and artists.

East End, London

Changes in perception:
Most younger generations in the East End value the area's cultural diversity, arts scene, and proximity to the financial district. Some older or ex-residents remember the area's manual work and pubs or cafes.

Economic transformation:
Glasgow diversified from industry (shipbuilding and docks) to a hub for education, culture, and retail.

Glasgow, Scotland

Interconnectedness:
Many younger generations are more connected to global cultural trends and digital technologies, and prefer urban living in neighbourhoods such as Glasgow's West End and cosmopolitan experiences.

Key terms Make sure you can write a definition for these key terms

intergenerational   segregation

6 Diverse places

6.9 Tensions and conflict

Changes that create challenges and opportunities

Changes to land use can benefit or disadvantage different groups of local people and impact their lived experience.

Action by	Opportunities	Challenges	Impacts on lived experience
Community groups e.g. construction of new places of worship	✅ Daily worship and activities	❌ Neighbours of other religions may feel out of place	• Helps residents of the religion to develop deeper geographical roots
Local governments e.g. low-traffic neighbourhoods (LTNs)	✅ Less air pollution ✅ Fewer accidents ✅ More active travel, which improves health of residents	❌ Some residents need to drive further to avoid LTNs to conduct business or visit relatives	• More pleasant at street level – less noisy, safer, more residents interacting
National governments e.g. 2020 Town and Country Planning Order, which fast-tracked the demolition and reconstruction of buildings	✅ Modern buildings replace older ones ✅ Jobs are created both in construction and within new buildings ✅ Newer and more energy efficient housing	❌ Loss of local landmarks ❌ Disruption to local businesses during construction	• Rapid changes in built environment can cause short-term disruption • Chance for neighbourhoods to reset their reputation
TNCs e.g. in the 1990s, major banks moved into the old dockland area of Canary Wharf, London	✅ Improved built environment ✅ Jobs, many with high pay ✅ Increased tax revenue can improve services and infrastructure for locals	❌ Most well-paid jobs did not go to local residents ❌ Few opportunities for ex-dock workers	• Many local residents feel separated from more affluent aspect of London

Tensions over the diversity of living spaces

Around 83% of the UK population live in urban areas. There are sometimes tensions over change and continuity in densely populated areas.

- Many long-term residents seek continuity, and have an emotional attachment to places connected to their personal history, families, and friends.

- Recent in-migrants often seek change to living space, such as culture-specific shops and services.

- Affluent newcomers invest in and renovate properties (gentrification), potentially displacing low-income residents.

- Cultural tensions may emerge over issues such as music, noise, religious practices, and different cultural practices.

- Housing shortages and homelessness contribute to tensions as various groups compete for limited resources. Perceptions of unfairness in allocation of social housing, for example, can lead to suspicion and discord.

Local governments often promote community cohesion initiatives to bridge gaps between residents with different lived experiences.

Diversity in Luton, Bedfordshire

Diverse town with a significant Asian population.

High demand for affordable homes (commuting distance to London) leads to competition between long-term residents and newcomers.

Debates over which cultural festivals should receive public funding.

Competition for the use of communal spaces can arise when different cultural or religious groups seek to utilise the same facilities.

Approximately 23.5% of residents do not have English as their first language. This can lead to misunderstandings and social isolation, hindering effective integration.

Some residents may be unfamiliar with the customs and traditions of other groups in Luton.

Tensions between groups have been stoked by extremists.

Changes to the built environment

New buildings, or changes to the use of existing structures, benefit members of some groups who may have more places to worship, meet, and receive communal support. This is especially valuable for some groups who experience **social exclusion** (such as refugees and asylum seekers) or discrimination.

London's Trocadero cinema plans to turn into a mosque

- The Metro cinema inside the Trocadero is an iconic building facing Piccadilly Circus. It closed in 2006.
- Around 15% of Westminster's residents are Muslim, but existing places of worship are full. Plans for a space for 1,000 worshippers were withdrawn in 2020 due to opposition.
- In July 2023, Westminster City Council approved plans for a Piccadilly Prayer Space for 390 worshippers, with spaces for other activities, including interfaith meetings and LGBTQIA+ talks.
- Groups who oppose the plan claim that they fear a threat to 'British culture'.

REVISION TIP

Check any new developments in the case studies you have learned about as you approach your exams, so you have the most up-to-date information to use in your answers.

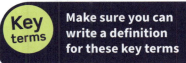

Key terms Make sure you can write a definition for these key terms

diversity   social exclusion

RETRIEVAL

Learn the answers to the questions below, then cover the answers with a piece of paper and write as many as you can. Check and repeat.

Questions

Answers

	Questions	Answers
1	What is the ratio of students to non-students in Manchester and Salford?	One to eight
2	What was the median age of residents of Hackney in 2021?	32
3	Describe two policies which have encouraged migration to the UK.	UK government promotion of large-scale immigration from countries that had been colonised by Britain (1940s to 1960s); free movement in the EU (1992 to 2021)
4	Name one UK policy which has discouraged migration.	One from: Brexit / points-based system of eligibility
5	International migrants tend to live in distinctive places, closely related to which two indicators?	Economic indicators and social indicators
6	Name two non-Christian religious festivals celebrated in Southall.	Two from: Diwali / Vaisakhi / Eid al-Fitr
7	Which generation is more likely to rent rather than buy houses?	Young adults
8	What is the local government policy known as an LTN?	Low-traffic neighbourhoods
9	Name three challenges that can cause tension over the diversity of living spaces.	Three from: long-term residents seeking continuity / recent in-migrants seeking change / gentrification / cultural tensions / housing shortages
10	Name one social feature which might be found in a diverse living space in an urban area.	One from: international grocery stores, and specialty shops and services / places of worship serving specific religious communities / leisure and cultural centres (e.g. community halls)

Put paper here

Previous questions

Now go back and use these questions to check your knowledge of previous topics.

Questions

Answers

	Questions	Answers
1	Give two reasons why some urban locations are perceived as undesirable.	Two from: crime rates / environmental quality / reputation / population characteristics
2	Name two sources of quantitative data that can affect perceptions of urban areas.	Two from: recorded crime / census data / surveys / Index of Multiple Deprivation (IMD)
3	Name two factors that can affect lived experience.	Two from: age / length of residence / levels of deprivation / ethnicity / gender / sexuality
4	Where are commuter villages often found?	Rural–urban fringe
5	Name two infrastructure challenges experienced by many rural areas.	Two from: narrow roads / unreliable internet / limited utilities

Put paper here

 KNOWLEDGE

6 Diverse places

6.10 Managing cultural and demographic issues

Assessment using measures of income

Income can be measured using the IMD dimension of income deprivation. This shows whether people are out of work or on low earnings. Household income includes wages and income from property and financial assets (but does not consider living costs).

On a regional level, in 2021, income levels were higher in south-east England, Cheshire, East Cumbria, Edinburgh, and east Scotland than in other parts of the UK. However, national patterns hide wide local differences, for example, in London.

Poverty

Low pay is a major cause of poverty.

Relative poverty is contextual. In the UK, households are below the poverty line if their income is 60% below median household income after housing costs for that year.

Poverty

Absolute poverty is measured against a set limit. People in absolute poverty are living below an income threshold that is considered necessary to meet basic needs, such as food, shelter, and clothing.

Assessment using measures of employment

- The IMD dimension of employment deprivation measures the percentage of working-age people who are economically inactive. That is, they are unable to work due to unemployment, sickness, disability, or care responsibilities.
- Economic inactivity is a major cause of poverty.
- Levels of economic inactivity vary for different ethnic groups within the UK.
- In 2023, unemployment was highest in Wales (5%), the West Midlands (5%) and London (4.9%)
- Unemployment was lowest in Northern Ireland (2.5%). But Northern Ireland had the highest percentage of economically inactive people (26%, compared to 20% for the UK).

Measuring social progress by reductions in inequality

Income inequality

Income inequality in the UK rose rapidly in the 1980s. It has since broadly levelled off.

Incomes in the East and South-west are rising, as these areas join the South-east and London in having much higher incomes than the rest of the UK. However, the high cost of housing means that high levels of inequality exist in London, with low homeownership rates.

Ethnic inequality

The hourly median pay gap between people of White British ethnicity and other ethnicities is shrinking, but there is still significant inequality.

REVISION TIP

Remember that different ethnic groups will have experienced different rates of social progress in recent decades.

 **KNOWLEDGE**

6 Diverse places

6.10 Managing cultural and demographic issues

Measuring social progress by improvements in health indicators

Health is closely linked to living and working conditions and inequities (unfair conditions of access) in power, money, and resources. These are the social determinants of health.

- The 2020 update to the 2010 Marmot Review found that, in the 2010s, life expectancy in England stalled, and the amount of time spent in poor health increased. Scotland has seen similar trends.

- People living in the north of England experience worse health outcomes than those in the south, but there are huge regional differences.

- Life expectancy is 8.6 years lower in the most deprived 10% of communities than in the least deprived communities.

- The UK government's widely questioned 2021 CRED (Commission on Race and Ethnic Disparities) report controversially stated that ethnicity was not the major driver of health inequality in the UK, but that deprivation was a key factor.

> **REVISION TIP**
>
> Try plotting case studies on a map of the UK to help you remember key facts.

Measuring integration and engagement

Political engagement

- In general, more deprived areas have lower levels of voting in the UK.

- Eligible citizens from marginalised ethnic backgrounds are less likely to be registered to vote than those of White British ethnicity.

- Higher voter turnout from diverse communities may suggest an increased sense of belonging and participation in the political process, reflecting positive **integration**.

- The requirement to show photo ID in order to vote in some UK elections may discriminate against some ethnic minority groups.

Community engagement

The creation of local community groups such as the multicultural 'Together as One' (*Aik Saath*) group in Slough, demonstrates integration and fosters cultural expression, networking, and support.

Levels of hate crime

- A reduction in hate crimes can signify progress in cultural integration. Lower levels of such crimes may indicate improved social cohesion and reduced tensions among different cultural groups.

- Public distaste for overt racism, and support for its victims, is growing.

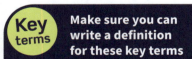 **Make sure you can write a definition for these key terms**

engagement   income inequality   integration

6.11 Managing change in urban communities

Contrasting views in Slough, Berkshire

Long-established migrant communities, including migrants from Wales and Scotland in the 1920s and 1930s, from Poland since 1945, and from India since the 1950s.

Migration due to proximity to jobs in London and at Heathrow Airport has led to pressure on houses and services.

Demographic and ethnic groups in Slough

Slough experienced tensions between groups of young Muslims, Hindus, and Sikhs in the 1980s and 1990s.

There is an absence of distinct ethnic enclaves, which can foster cohesion and increase trust.

Local strategies to resolve issues

Together as One (formerly *Aik Saath*) is a community organisation founded in 1998, working to resolve issues related to **community cohesion** in Slough. Strategies include:

- youth engagement programmes, including workshops, leadership training, and mentoring initiatives, to foster a sense of belonging and shared purpose
- organised cultural and interfaith events that celebrate the diversity of Slough
- community workshops on issues like racism and social inclusion, allowing the exchange of ideas
- youth-led initiatives, including community projects and campaigns
- collaboration with other agencies (local government, schools, businesses, and other community organisations) to gain funding and access to decision-makers.

REVISION TIP
Find out about a cross-community initiative and compare it to 'Together as One'.

National strategies to resolve issues

- The Equality Act 2010 prohibits discrimination, including on the basis of race and ethnicity.
- In 2021, a CRED report recommended building trust, promoting fairness, creating agency, and achieving inclusivity. The report found that discrimination was falling, but this finding has been questioned and challenged by a number of individuals and organisations.
- UK legislation has been updated to address hate crimes.

REVISION TIP
Remember that success may be judged using short- and long-term criteria.

Assessing success in Slough

Economic success?	Social and demographic success?	Environmental success?
- Studies have found that Slough is one of the most productive towns in the UK. - Many high-tech TNCs have offices in Slough. - However, Slough Borough Council effectively declared bankruptcy in 2021.	- Slough Borough Council has promoted housing, culture, and sports. - There is still great need for affordable housing. - Crime rates have risen. - Tensions in the 1980s and 1990s have been diffused, and many see Slough as a successful diverse community.	- Residents of Slough can access the Berkshire countryside. - Slough experiences high levels of air pollution at street level (many commuters drive) and noise and air pollution from air traffic related to Heathrow. - Slough is outside of the Greater London ULEZ so will not benefit from the related reductions in air pollution. However, electric car infrastructure is increasing.

6   Diverse places

6.11   Managing change in urban communities

Contrasting success criteria

Different stakeholders will assess success using contrasting criteria, depending on their role, the meaning of the place for them, and the actual and perceived impact of changes to the place.

Stakeholder	Criteria
National government	Higher global status for UK, tourism income (Heathrow), tax revenue (e.g. TNC HQs)
Local council (Slough Borough Council)	Housing, employment, skills, transport, sports amenities
Non-profit organisations (e.g. Together as One)	Community cohesion, lower street crime, fewer hate crimes, long-term financial sustainability of schemes
Local businesses	Trade, infrastructure, continuity
Local communities	Housing, employment, skills, transport, sports amenities

Impact of change

The physical fabric of Slough changed significantly in the building boom of the 1960s, but developments since then have had less impact on the built environment. The lived experience will depend on people's personal context, including their daily encounters and their long-term social, cultural, and economic circumstances. The reality of life for residents may differ from the image cultivated by those who want to show off its successes.

Interest groups are heterogenous (varied). Residents can be split into subcategories (which intersect), including:

- Landlords who have turned single family houses into houses of multiple occupancy (HMOs) to respond to increases in numbers of residents.
- Younger residents who generally welcome economic, social, and cultural revitalisation.
- Older residents who tend to value local interactions and have developed deep roots (e.g. at cafes and pubs).
- Community workers and volunteers who notice less outward discrimination.
- People with visible and less visible disabilities who are sometimes overlooked when assessing success.
- People of different ethnicities may have different points of view.

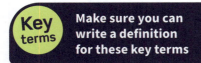 **Make sure you can write a definition for these key terms**

community cohesion

6.12 Managing change in rural communities

Perceptions of rural areas

- Rural areas are often portrayed as being peaceful, clean, traditional, stable, safe, and prosperous.
- They are also often portrayed as being 'old' and 'white'. Rural-to-urban migration (especially of the young and highly educated) depletes some areas of young adults.
- The perception is of fewer relevant economic opportunities, fewer multicultural facilities, and less acceptance of change. This can mean that people from marginalised ethnicities may be less likely to choose to live in rural areas.

North Antrim coast, Northern Ireland

- The coastline is famous for the Giant's Causeway (UNESCO World Heritage site), golf, and its scenery.
- Around 40% of the population is over 45 (compared to 39% for Northern Ireland).
- Approximately 1% of the population are from marginalised ethnic backgrounds (compared with 3.4% for Northern Ireland, and 18.3% for England and Wales).
- Challenges in the area include remoteness, the legacy of Northern Ireland's 'Troubles', farm profitability, and coastal erosion.

Strategies to resolve issues

Some organisations in Northern Ireland have worked together to help people from marginalised ethnic backgrounds, including:

- Armagh Traveller Support Group (ATTS), which supports people from Irish and Roma Traveller communities.
- Empowering Refugees and Newcomers Organisation (ERANO).
- These and other groups participated in a diversity conference held in April 2023.

REVISION TIP
Remember that these strategies often overlap.

- Sports-led: Portstewart golf course and other golf tourism attractions.
- Travel-led: promotion of the causeway coastal route.
- Heritage-led: Giant's Causeway, including £18m visitor centre (2013) and castles.

Strategies to resolve economic and social challenges

- Diversification-led: helping farms diversify into adventure-led tourism.
- Food-led: local food at Causeway market, Coleraine; branding of Menu of Moyle; whiskey.

Opposition to strategies

The need for income from tourism	vs	The loss of local character as the hospitality industry increases investment
The need for local economic revitalisation	vs	Longstanding attachments to rural ways of life (e.g. cattle farming) regardless of their profitability
The demand for new golf resorts and the associated employment	vs	The loss of ecosystems (dunes and other coastal habitats)
The improved quality of the Causeway Coastal Route for drivers and coach companies	vs	Congestion and air/noise/visual pollution for all local residents, especially those with breathing difficulties (up to 85 coaches a day visit the Giant's Causeway)

 **KNOWLEDGE**

6 Diverse places

6.12 Managing change in rural communities

Assessing rural regeneration on the North Antrim coast

Economic success?

- Tourism is seasonal and fickle. Travel restrictions (e.g. Covid-19), fashions, and economic trends have impacted areas.
- The failure of Bushmills Dunes golf resort reduces potential for job and income generation, especially for young people.
- Around 25% of employment is in the public sector (higher than the Northern Ireland average). Does this offer less economic value?
- Farms that have diversified and draw income from produce and day visitors are thriving.

Social and demographic success?

- Around 71% of working-age adults are economically active. This is lower than for Northern Ireland as a whole (73%). This shows there is scope for more regeneration. Jobs in farming and manufacturing are falling.
- Educational attainment is in line with the rest of Northern Ireland.
- Portrush has a young age profile, aided by pubs, nightclubs, and restaurants.
- Local strategies have increased tolerance and improved acceptance of cultural diversity.

Environmental success?

- The failure of Bushmills Dunes golf resort, opposed by environmentalists, shows that the natural environment has been prioritised.
- The Giant's Causeway area is well protected (UNESCO status, Area of Outstanding Natural Beauty, over 90% of land owned by the National Trust), but coastal paths are threatened by erosion.
- The Giant's Causeway visitor centre, built unobtrusively into a hillside.

Contrasting success criteria

Stakeholder	Criteria
UNESCO	Maintain globally important geological and natural heritage
National government	Higher global status for the UK, tourism income, tax revenue
Devolved government (Northern Ireland)	Higher status, reduce rural inequalities, tax revenue, many tourists will spend time at other attractions in Northern Ireland (e.g. Titanic Museum, Belfast)
Local council (County Antrim)	Employment, skills, transport
Causeway Coast and Glens Borough Council	Encourage visitors to stay for more than a day trip, circulating more money in the local economy
Developers	Profit, status, reputation
Local businesses	Trade, infrastructure, continuity
Local communities	Housing, employment, skills, transport, sports amenities

Impact of change

Lived experience will depend on personal context, but the reality of life for residents may differ from the image cultivated by those who want to show off its successes.

- Younger residents generally welcome the job opportunities presented by tourism and diversification.
- Older residents tend to value local interactions.

- People of marginalised ethnicity may well lose out on important cultural and religious hubs.
- Small business owners and residents are pleased in peak tourist times, but experience quiet periods too.

RETRIEVAL

Learn the answers to the questions below, then cover the answers with a piece of paper and write as many as you can. Check and repeat.

Questions | Answers

#	Questions	Answers
1	How can household income be measured?	Using the IMD income deprivation dimension
2	What is absolute poverty?	Living on an income below the threshold considered necessary to meet basic needs
3	Which report claimed that deprivation and geography are more important factors than ethnicity in driving health inequalities?	The UK government's 2021 CRED (Commission on Race and Ethnic Disparities) report
4	Name three Together as One initiatives.	Three from: youth engagement programs / cultural and interfaith events / community workshops / youth-led initiatives / collaboration with other agencies
5	Name two UK government strategies aimed at reducing ethnic tensions	Two from: Equality Act / CRED report / hate-crime legislation
6	Why do people of marginalised ethnicities tend not to live in rural areas in the UK?	Perception of fewer relevant economic opportunities, fewer multicultural facilities, and less acceptance of change
7	What is the approximate percentage of the North Antrim coast population who are from marginalised ethnic backgrounds?	1%
8	Name one group or strategy aimed at assisting people from ethnically marginalised backgrounds in Northern Ireland.	One from: Armagh Traveller Support Group / Empowering Refugees and Newcomers Organisation / 2023 diversity conference
9	What industry tends to employ many young people on the Antrim Coast?	Tourism or hospitality

Put paper here

Previous questions

Now go back and use these questions to check your knowledge of previous topics.

Questions | Answers

#	Questions	Answers
1	Name one UK policy which has discouraged migration.	One from: Brexit / points-based system of eligibility
2	Which generation is more likely to rent rather than buy houses?	Young adults
3	What is the local government policy known as an LTN?	Low-traffic neighbourhoods
4	Name three challenges that can cause tension over the diversity of living spaces.	Three from: long-term residents seeking continuity / recent in-migrants seeking change / gentrification / cultural tensions / housing shortages
5	Which UK region has the youngest average age?	London (35.9 years)

Put paper here

PRACTICE

1   Study **Figure 1**. Suggest **one** reason why the speaker may have different
    perceptions of different parts of London.   **(3)**

> "Yeah, so I think of South London being really my place, my roots. I'm South
> London. I sound like it. I love it. My husband is South London. I'm really proud
> of London because it's so multicultural. And so diverse. When I go away to
> certain places, I feel certain in South London. I remember moving and I moved
> to the outskirts of London, still in London, but the outskirts. That for me was
> a little bit scary. My first instinct was, are there going to be people like me?
> Am I going to see people like me? Oh don't, don't laugh. But when I saw like, a
> Black or Asian face or someone that was different, I got excited - it was like oh,
> it's okay, we're gonna be okay."

▲ *Figure 1 An excerpt from an interview with Lisa, a resident of South London
with Jamaican and Guyanese heritage, recorded by The Voices Project.*

2   Suggest **one** reason why ethnic diversity varies between settlements.   **(3)**

3   Suggest **one** reason why fertility rates vary between places.   **(3)**

> **EXAM TIP**
>
> Three-mark questions
> require you to give one
> reason, then make two
> points of development.
> Think of these as 'reason,
> so, so' answers.

4   Study **Figure 2**. Suggest **one** reason why some suburban areas are perceived
    differently by contrasting demographic groups.   **(3)**

> Kenyon Park* has been used by groups of young people, some of whom are
> drinking alcohol and using drugs.
>
> Unfortunately, incidents of anti-social behaviour from such groups are
> growing again. Loud music has been played late at night, and litter associated
> with drinking and drug-taking has been found in the more isolated places in
> the park, especially in the bottom copse and on the wooded slope.

> **EXAM TIP**
>
> Ensure you read the
> question carefully – is it
> asking about urban areas,
> rural areas, or both?

*Name, and some other details, have been changed

▲ *Figure 2 Excerpt from website of Friends of Kenyon Park* (in the suburbs of a
large UK city)*

5   Suggest **one** reason why some rural locations may be seen as undesirable
    by outsiders.   **(3)**

6   Explain why some urban areas are perceived to be
    undesirable or threatening.   **(6)**

7   Explain how changes to the built environment can bring benefits to some
    groups but provoke hostility from others.   **(6)**

8 Study **Figure 3**. Explain how different people may have varying perceptions of this area. **(6)**

▲ **Figure 3** *The village of Conistone, Upper Wharfedale, Yorkshire Dales*

9 Explain **one** way in which fertility rates are changing the cultural characteristics of places. **(6)**

10 Explain how the media can provide contrasting evidence about the image different people have of your chosen local place. **(6)**

11 Explain why rural areas are often seen as idyllic. **(6)**

12 Explain why population density varies within the rural–urban continuum. **(6)**

13 'International migration is the main reason why the populations of places vary'. Evaluate this statement. **(20)**

14 Evaluate the extent to which international and global influences have shaped the characteristics of your local place. **(20)**

15 Evaluate the view that levels of segregation within places reflect cultural, economic and social variation and change over time. **(20)**

16 Evaluate the range of ways which can be used to determine how residents view your local place. **(20)**

7 The water cycle and water insecurity

7.1 Hydrological cycle

The global hydrological cycle's operation as a closed system

- There is a finite amount of water on Earth which is moved through stages of the hydrological cycle as ice, liquid water, or water vapour.

- It is a closed system, meaning no external **inputs** or **outputs** (no water lost or gained). Instead, there is a series of **flows** (**fluxes**) and **stores** of water on Earth.

Processes operating in the cycle include evaporation, condensation, precipitation, evapotranspiration, and surface and groundwater flows.

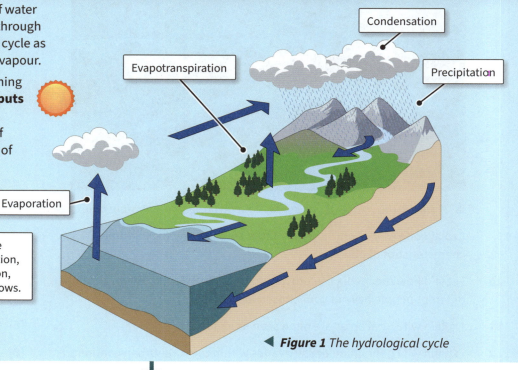

◀ *Figure 1* *The hydrological cycle*

Drivers of the hydrological cycle

- Solar energy: heat from the Sun melts ice and causes evaporation. When energy is removed (temperature falls), water vapour condenses and water freezes, forming ice.

- Gravitational potential energy: gravity pulls water downwards, over or through the land, and into the oceans.

Water stores

- There are five main stores:
  1. Water stored as liquid in the oceans. This is the most important store.
  2. As glacial ice, snow, or permafrost in the **cryosphere**.
  3. As groundwater in rocks below ground.
  4. On land as surface water in lakes and rivers. This makes up a very small proportion of total water but it is the most accessible for human use.
  5. Water vapour in the atmosphere. The amount here is relatively small and it is precipitated quickly.
  6. In living organisms in the **biosphere**.

- Stores can be depleted or enlarged depending on the balance of their inputs and outputs.

- Climate change can alter this balance. An ice age will cause the cryosphere to grow significantly and reduce the ocean store.

Store	Proportion of total water
Oceans	96.5%
Cryosphere	1.7%
Groundwater	1.7%
Surface water	0.01%
Atmosphere	0.001%
Biosphere	0.000,01%

Annual water fluxes

Water moves between stores. The rate at which a flow occurs is a flux.

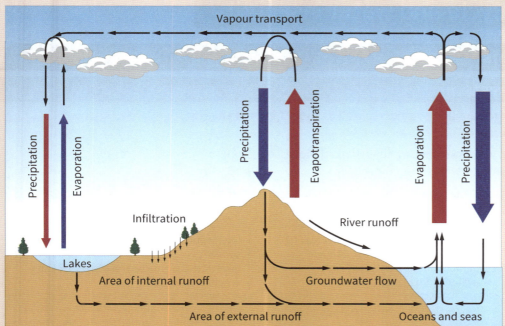

▲ **Figure 2** *Proportional flows (fluxes) in the hydrological system*

Flux movement	Process	Flux (km³ per year)
Ocean and atmosphere	Evaporation	502,800
	Precipitation	458,000
Atmosphere and land	Evaporation	74,200
	Precipitation	119,000
Land to ocean	Surface runoff and groundwater flow	44,800

Limited water availability

- Water for human use is very limited, with freshwater making up only 2.5% of Earth's total water. Most of this is inaccessible or locked up in the cryosphere.

- River water, which derives from precipitation or glacial melt, constitutes a tiny proportion of total water at 0.007%.

Residence times

- Stores keep water for different lengths of time. These are known as **residence times**.
  - Longer residence times may make freshwater inaccessible or difficult to access.
  - Most residence times are short, with that of atmospheric water vapour being only days.
  - Groundwater, oceans, and ice caps can store water for millennia.
- Some freshwater stored deep in the ground fell as rain millennia ago and will not be restored if used by people. This means it is non-renewable **fossil water**. The Nubian sandstone aquifer under the Sahara Desert is an example.

Key terms Make sure you can write a definition for these key terms

biosphere   cryosphere   flow   flux   fossil water   input
output   residence time   store

7 The water cycle and water insecurity

7.2 Drainage basins

The drainage basin system

- The **drainage basin** is the area drained by a river and all its tributaries, the edge of which is marked by its watershed.
- The hydrological cycle is an **open system** containing linked processes of inputs, flows (throughputs), and outputs, with stores in between.
- Inputs and outputs can vary and the total amount of water in a basin is affected by external factors.

Saturated overland flow: water runs over the land and into rivers when the ground is already full of water (**saturated**) during a rainfall event.

Soil moisture storage: the water held within permeable soil.

Percolation: the movement of water, under the influence of gravity, from the soil into permeable rock below.

Groundwater store: water held in **permeable** rocks underground, known as an **aquifer**.

Infiltration (flow): the movement of water into the soil.

Groundwater flow: the slow movement of water through underground rocks.

Precipitation (input): rainfall, snow, or hail are very important as the only input. If input exceeds the outputs, then the basin will be in surplus and this could lead to floods. The pattern of precipitation depends upon the season and long-term cycles of climate change.

Evapotranspiration (output): water leaving the drainage basin as vapour through evaporation (of surface water) and transpiration (of plants)

Interception storage: water held on the surface of vegetation after interception. Water taken into the structure of plants becomes vegetation storage.

Interception (flow): vegetation catches precipitation and transfers it elsewhere.

Surface storage: water held on the surface in ponds, puddles, and lakes, and as ice on mountains and in glaciers.

Channel storage: the water held within rivers.

Channel runoff: the loss of water from a river into the sea.

Throughflow: water movement through the soil and into a river.

◀ **Figure 1** The drainage basin open system contains inputs, flows, stores, and outputs

> **REVISION TIP**
>
> The balance of inputs and outputs is a budget. If total outputs exceed inputs, the amount of water in a drainage basin is in deficit.

Types of precipitation

Orographic rainfall: where high relief forces air masses upwards causing water vapour to condense. The areas inland from a mountain range may experience less rainfall as the atmospheric moisture has already been lost over the mountains. This rain-shadow effect means the leeward side of a mountain range can be dry.

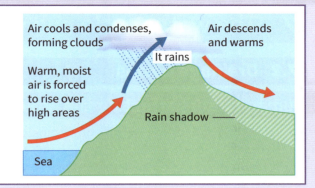

Frontal rainfall: where colder air meets warmer, moisture-laden air, leading to condensation of water vapour along bands of weather fronts.

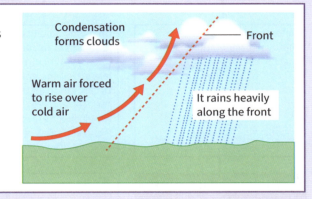

Convectional rainfall: warm, moist air rises rapidly in tropical areas or, in the UK, during late summer. It cools rapidly and releases the moisture as thunderstorms and heavy rainfall.

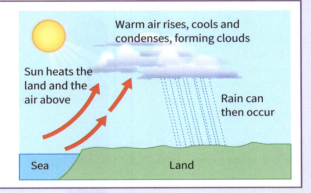

Physical factors affecting the drainage basin

Physical factors control the relative importance of inputs, flows, and outputs.

Climate: the amount of precipitation determines inputs, which feed into stores and flows and, ultimately, outputs. Areas with seasonal changes may have complex patterns across the year. Climate also determines vegetation quantity and type.

Vegetation: greater amounts will intercept more precipitation, store water as vegetation storage, and encourage more evapotranspiration. Roots will also increase infiltration rates.

Soils: loose, permeable soils will encourage infiltration and reduce surface runoff.

Physical factors

Geology: the permeability of rocks controls the amount of percolation, groundwater storage, and flow. Impermeable rocks like granite increase the amount of water on the surface and in soil.

Relief: higher land will experience more rainfall and steeper slopes encourage more overland flow.

7.2 Drainage basins

Human factors affecting the drainage basin

Human factors can accelerate or slow down processes.

Human factors

Deforestation: the removal of vegetation means less interception, vegetation storage, and evapotranspiration, which can lead to greatly increased overland flow, encouraging flooding.

Changing land use: conversion to farmland or **urbanisation** has huge impacts. Agricultural fields intercept less water and encourage overland flow. Urbanisation increases impermeable surfaces and creates drains, moving water directly into rivers, leading to increased rates of overland flow and channel flow.

Creating new water storage reservoirs: the creation of reservoirs increases surface storage and accelerates evaporation, whilst dams slow channel flow, especially upstream. Water can also be more easily removed for human use.

Abstraction: pumping water from the ground severely reduces groundwater storage in an aquifer and lowers the water table, which can lead to rivers drying up completely. About 40% of London's water comes from chalk aquifers.

Deforestation in Amazonia

Over 20% of the Amazon rainforest has been deforested, with 6% more severely degraded for logging and to make space for cattle ranches and farming. This has caused major disruption to the drainage basin, to the extent that local climates have been altered. In some areas:

- Interception has been reduced, causing a significant decrease in evapotranspiration. This may also reduce precipitation in the long term.
- Overland flow has increased, taking soil and silt into rivers and washing away nutrients.

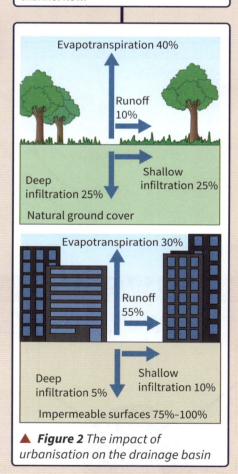

▲ *Figure 2 The impact of urbanisation on the drainage basin*

Key terms

Make sure you can write a definition for these key terms

abstraction   aquifer   convectional rainfall   drainage basin
frontal rainfall   open system   orographic rainfall
permeable   saturated   urbanisation

7.3 Water budgets

Water balance

- The annual balance of water between inputs and outputs is a **water budget**.
- A negative budget means outputs are greater than inputs, leading to a negative balance of water in a system – water is lost over time and in **deficit**. A positive balance would see a **surplus** of water.
- This can be applied at a global, regional, or local scale and varies with the seasons in most places:
  - Tropical temperate climates are hot and wet due to their location on the Equator where trade winds create a low-pressure zone, the Intertropical Convergence Zone (ICTZ), and have a high flux of exchange between inputs and outputs.
  - Polar climates are cold and dry due to high pressure and have a low flux between inputs and outputs.

Water balance equation

An equation can be used to understand the balance of water in a drainage basin:

$$P = E + Q + \Delta S$$

This states that the input, precipitation (P), should balance with the outputs of evapotranspiration (E) and runoff (Q) when changes in water held as soil moisture and groundwater storage within the basin (ΔS) are accounted for.

REVISION TIP

Try to understand the idea of the balance rather than focussing on a formula involving numbers.

Water budget graphs on a local scale

B Potential evapotranspiration rises above precipitation in the summer (output above input), meaning the store of water in the soil is being used up (**utilised**).

C A point comes when soil moisture has been fully used.

D The drainage basin is now without soil moisture and potential evapotranspiration is still above precipitation, leading to a water deficit.

E In the autumn, precipitation rises above potential evapotranspiration again and the soil regains its lost moisture (**recharge**).

A Precipitation is above potential evapotranspiration (input above output), meaning a surplus of water is available.

F/G When the soil has completely regained its lost moisture (field capacity, F) the water budget will again be in surplus (G).

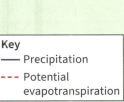

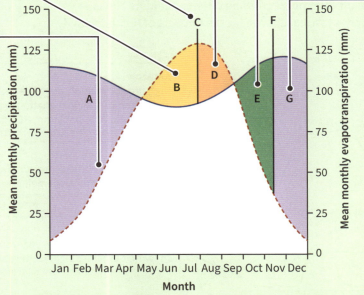

Key
— Precipitation
--- Potential evapotranspiration

▲ *Figure 1 The change in the balance of inputs (precipitation) and outputs (potential evapotranspiration) on soil moisture over a year can be shown in a* ***water budget graph*** *for an area*

7.3 Water budgets

Analysing water budget graphs

Individual graphs can also be analysed to understand the need for water management.

▶ **Figure 2** In some areas, precipitation always lies above potential evapotranspiration, meaning the basin is always in surplus (e.g. Tokyo, Japan)

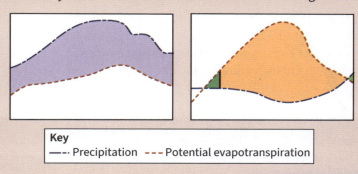

Key
—·— Precipitation   --- Potential evapotranspiration

◀ **Figure 3** In other areas, there is only a very short period where precipitation exceeds potential evapotranspiration, leading to a short recharge period. This is quickly utilised before the budget returns to deficit for most of the year (e.g. Baghdad, Iraq)

 REVISION TIP

Water budget graphs can be used when planning for possible deficits (droughts) or surpluses (floods).

River regimes

- A **river regime** indicates the annual variation of **discharge** of a river. It can be used to analyse the patterns of flow over the year and the impact of this on human activity.

- A river's regime depends on the climate, geology, and soils of the drainage basin.

- Some vary dramatically across the year, whilst others remain constant.

- Complex regimes may result from rivers crossing different climate zones or types of geology, significant seasonal change, or the impacts of human activity such as abstraction.

 REVISION TIP

These three river regimes can be used for comparison. Make sure you appreciate the logarithmic scale when analysing the changes across the year.

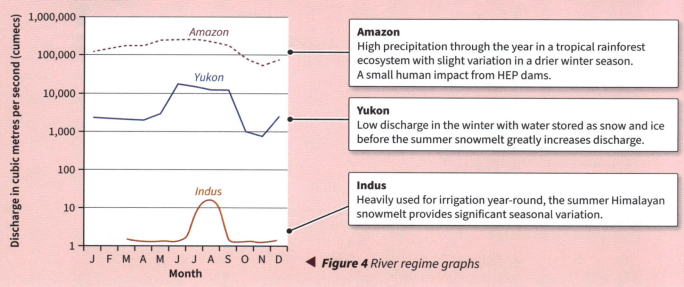

Amazon
High precipitation through the year in a tropical rainforest ecosystem with slight variation in a drier winter season. A small human impact from HEP dams.

Yukon
Low discharge in the winter with water stored as snow and ice before the summer snowmelt greatly increases discharge.

Indus
Heavily used for irrigation year-round, the summer Himalayan snowmelt provides significant seasonal variation.

◀ **Figure 4** River regime graphs

Key terms — Make sure you can write a definition for these key terms

deficit   discharge   recharge   river regime   surplus   utilised
water budget   water budget graph

7.4 Storm hydrographs and river regimes

Storm hydrographs

- A storm **hydrograph** is a graphical representation of rainfall and discharge following a single rainfall event rather than discharge over a whole year (see river regimes).
- Water from rainfall reaches the river through various flows.
- Flows are measured in **cumecs** (cubic metres per second) at a single location.
- They can be used to calculate **lag time** and to monitor flood risk.

REVISION TIP

Lag time is the most important aspect, so prioritise this for retrieval.

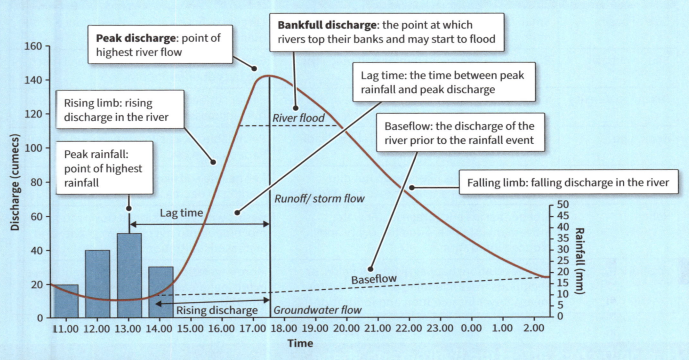

▲ *Figure 1 Features of a storm hydrograph*

Hydrograph shapes

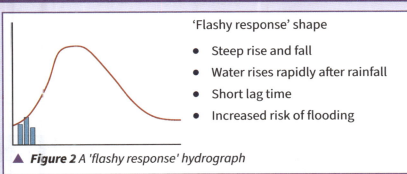

'Flashy response' shape

- Steep rise and fall
- Water rises rapidly after rainfall
- Short lag time
- Increased risk of flooding

▲ *Figure 2 A 'flashy response' hydrograph*

REVISION TIP

Remember that real storm hydrographs can be complex and have multiple peaks of rainfall and discharge.

'Gentle response' shape

- Gentle rise and fall
- Water rises more slowly after rainfall
- Longer lag time
- Reduced risk of flooding

▲ *Figure 3 A 'gentle response' hydrograph*

7 The water cycle and water insecurity

7.4 Storm hydrographs and river regimes

Physical factors affecting hydrograph shape

Physical factors	Flashy shape	Gentle shape
Size of basin	Small basins have a shorter distance to the main river channel	Larger basins have a longer distance to the main river channel
Shape of basin	Rounder basins mean rainfall will reach the river at a similar time	Elongated basins mean rainfall will reach the river at different times
Drainage density	More surface streams leading to the main river reduces lag time	Fewer surface streams leading to the main river increases lag time
Rock type	Impermeable rocks lower infiltration, leading to increased overland flow	Permeable rocks allow infiltration, leading to decreased overland flow
Soil	Impermeable soils like clay do not allow infiltration, leading to increased overland flow	Permeable soil like sand allow infiltration, leading to decreased overland flow
Relief	Steeper basins encourage faster water movement into rivers due to gravity and reduced infiltration	Basins with gentle slopes slow water movement into rivers due to gravity and increased infiltration
Vegetation	Less vegetation, bare agricultural fields, or deciduous trees shedding leaves in winter means lower interception and infiltration	More vegetation, planted crops, and trees in full leaf means more interception and infiltration

Human factors affecting hydrograph shape

Human factors	Flashy shape	Gentle shape
Land use	• Bare fields or crops like maize increase overland flow • Ploughing downhill discourages infiltration	• Planted fields of crops like wheat decrease overland flow • Contour ploughing encourages infiltration
Urbanisation	• More concrete and tarmac increase overland flow • Sewers and drains reduce lag time	• Afforestation increases interception and infiltration

Players: the role of planners

- Local and national governments control planning policy for housing and land use.
- A key part of planning today is the use of sustainable drainage systems (SuDS), which aim to reduce overland flow and flooding.
- Recent floods have been attributed to poor planning of new housing developments by developers.
- In the UK, flood risk is managed by the Environment Agency.

Key terms — Make sure you can write a definition for these key terms

bankfull discharge   cumec   drainage density   hydrograph
lag time   peak discharge

RETRIEVAL

Learn the answers to the questions below, then cover the answers with a piece of paper and write as many as you can. Check and repeat.

Questions

		Answers
1	Define a closed system.	No external input or output (nothing is lost or gained)
2	What two factors drive the hydrological cycle?	Solar energy and gravitational potential energy
3	Give two key stores in the hydrological cycle.	Two from: oceans / cryosphere / atmosphere / groundwater / surface water
4	What store has the longest residence time?	Groundwater
5	Give two types of rainfall.	Two from: orographic / frontal / convectional
6	Under what conditions can water be stored in an aquifer?	If the rock is permeable
7	What elements of the drainage basin system would be affected by a change in precipitation?	All of them
8	What human factors could increase overland flow?	Deforestation, changing land use, urbanisation
9	If inputs are higher than outputs, the water budget is in surplus. True or false?	True
10	Which component of the water balance equation is missing? $P = E + \rule{2cm}{0.4pt} + \Delta S$	Q (meaning runoff)
11	Define the term 'recharge'.	When soil regains lost moisture because precipitation is higher than potential evapotranspiration
12	Name one physical factor affecting hydrograph shape.	One from: size of basin / shape of basin / drainage density / rock type / soil / relief / vegetation
13	What two variables are represented on a hydrograph?	Rainfall and discharge
14	Name one human factor affecting hydrograph shape.	One from: land use / urbanisation
15	Name one player who may use a hydrograph for planning.	One from: local or national government / developers / the Environment Agency

Put paper here

7 The water cycle and water insecurity

7.5 Drought

The nature of drought

- Drought is a complex process related to the absence of water.
- It is seen as a chronic hazard as it is a creeping phenomenon which can be very long-term.
- It can be divided into four types which can be seen as a succession:
  1. **Meteorological drought**: when weather patterns lead to a short-term precipitation deficit.
  2. Hydrological drought: when the drainage basin reacts to this deficit with low water levels.
  3. Agricultural drought: when crops become affected.
  4. Socioeconomic drought: when people are affected through demand for goods exceeding supply because of the drought.

REVISION TIP

A dry area is not necessarily in drought. It depends whether precipitation is consistently below the average expected amount.

REVISION TIP

The four types of drought in sequence can be retrieved using a mnemonic: MHAS.

The physical causes of drought

Precipitation deficit

Continued low precipitation compared to the expected amount will affect the water budget.

- Short-term deficits: High pressure areas, known as **anticyclones**, can lead to weeks without precipitation, as low pressure belts containing moisture are forced around sinking air.
- Long-term cycles: Analysis of long-term climate data shows that some areas experience low precipitation rates across decades or centuries in natural climatic cycles. The frequency and extent of these oscillations are suggested to be affected by a natural phenomenon known as ENSO (the El Niño Southern Oscillation) which may be intensified by global warming (see 7.7).

Hydrological causes

Following a precipitation deficit, a drainage basin may rely upon its stores for water to flow. Once the stores are fully utilised, the drainage basin will be in hydrological drought, especially if the stores have been previously depleted or are inaccessible.

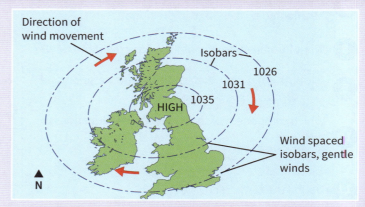

▲ **Figure 1** *A synoptic map showing a blocking anticyclone across the UK*

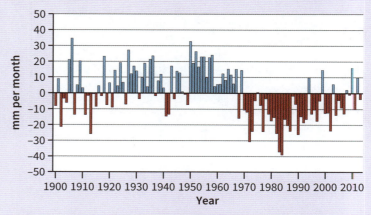

▲ **Figure 2** *A rainfall index of the Sahel region in Africa shows a clear long-term cycle of rainfall interspersed with shorter cycles*

Human contribution to drought

Over-abstraction

The removal of water from aquifers (bore holes and wells), rivers, and reservoirs for irrigation and industrial and domestic use is unsustainable if not replaced by precipitation. This is especially significant in dry areas with low or unpredictable precipitation.

Desertification

The removal of vegetation in semi-desert areas can lead to the land becoming desert due to increased erosion of soil by the wind, which further reduces the ability of remaining vegetation to grow.

Drought in Australia

- The 'Big Dry' drought lasted from 1996 to 2012 and affected the Murray–Darling river, with rainfall up to 20% lower than expected for this period.
- Over–abstraction for rapidly expanding urban areas and unrestricted irrigation, which accounted for 70% of total water usage, significantly contributed to impacts from climate change and ENSO cycles, increasing the severity of the drought.

Drought in the Sahel

- The Sahel region in Africa experienced 30% lower precipitation than expected from 1970 to 2010. The impacts of this were made worse by human activity led by poverty.
- High population growth (doubling every 25 years) resulted in overgrazing of vegetation by herded animals, overcultivation of crops, and high demand for firewood, leading to the loss of topsoil and the land slowly becoming desert.

The impacts of drought on ecosystems

Drought removes the amount of water available for ecosystems to function and reduces their resilience to future shocks. It may also lead to **feedback loops** or a **tipping point**, which can make the nature and extent of the drought worse.

Feedback loops and tipping points

Feedback is where an impact can either reinforce the original process (positive) to make the impact worse, or reduce it (negative). If positive feedback continues, a system can cross a threshold, known as a tipping point, leading to irreversible change and the system reaching a new, stable but changed, state.

Forest ecosystems

- Drought causes **forest stress**, meaning trees shed leaves and die.
- This leads to reduced evapotranspiration, which can reduce precipitation levels even further (positive feedback).
- Forest fires can also break out much more easily in the dry conditions, reducing vegetation even further.
- If a tipping point is reached, the ecosystem will change into a new state more suited to the drier conditions, e.g. tropical rainforest in the Amazon changing to savannah.

Wetland ecosystems

- Drought reduces the surface water store that makes up freshwater wetlands. Causes may relate to reductions in seasonal rains, such as those affecting the Pantanal, Brazil.
- The wetland dries out with huge impacts on all insects, fish, birds, and mammals dependent upon the biodiverse ecosystem, which may become prone to fires.
- It will also lose its function as a filter for pollutants and sediment, further affecting ecosystems.

> **REVISION TIP**
> Feedback and tipping points are core concepts for all physical geography. Use your retrieval practise to apply them to this topic.

Key terms
Make sure you can write a definition for these key terms

anticyclone   desertification   ENSO   feedback loops
forest stress   meteorological drought   overcultivation
overgrazing   tipping point

7 The water cycle and water insecurity

7.6 Flooding

The nature of floods

Floods occur when there is a surplus of water in a system, as inputs exceed outputs and stores exceed their capacity:

- Rivers may overtop their banks following heavy or prolonged rainfall events or due to snowmelt.

- Human activity can also increase the risks of flooding through land use changes and poor river management.

REVISION TIP

Try to make the link to systems terminology when revising flooding.

Physical causes of flooding

Depressions:
- Low-pressure frontal systems (**depressions**) bring especially heavy and prolonged rainfall, although any storm can lead to flooding.
- The **jet stream** in the upper atmosphere may convey a sequence of these systems to the same locations, meaning a succession of heavy rainstorms.

Intense storms:
Intense rainfall events may lead to saturation of soils and short-term surface flooding as infiltration capacity is exceeded.

Physical causes of flooding

Monsoon:
- A seasonal change in local climate of some areas may lead to a distinct rainy period in a year known as a **monsoon**.
- The extremely heavy rainfall of the India and South-East Asia monsoon, between April and September, falls onto previously very dry ground, commonly leading to flooding.

Snowmelt:
- Rapid snowmelt can overwhelm rivers, especially if large areas of snow melts at the same time and cannot to soak into frozen soil.
- In colder climates, this is a seasonal event flooding large areas, and can be made worse by ice blocking rivers.

Human actions that exacerbate flooding

Changing land use

- The conversion of grasslands and forests to farmland reduces interception and storage, and reduces the lag time into rivers by increasing overland flow.
- Bare winter fields may also encourage soil to silt up rivers, reducing the storage capacity of river channels.
- Urbanisation increases impermeable surfaces and channels water directly into rivers through drains, which may increase peak discharge.

Mismanagement of rivers

- River tributaries may have been straightened (**realigned**) from the original meandering course, meaning water enters the main channel more quickly.
- Rivers may also have been **channelised** with concrete embankments, speeding the flow, which may affect flood risk in different areas.

The damage from flooding

Environmental damage	Socio-economic damage
• Wildlife can be killed, with a loss of biodiversity. • Riverbanks are eroded and sediment laid down on floodplains, altering the river system. • Pollutants and contamination can be spread (e.g. oil, pesticides, and heavy metals). • Ecosystems may change following floods.	• A direct risk to people with 6000 killed on average per year in the last 20 years. • Homes and businesses are inundated, causing huge repair costs and disrupting economic activity. • Devastation of agricultural land in the short-term, especially key for subsistence farmers. • Waterborne diseases such as cholera may spread in countries with no direct water supply and limited healthcare.

The Cumbrian floods, UK, 2015

Background and causes

- Storm Desmond, an extratropical cyclone, produced exceptionally high rainfall across Cumbria, exceeding 300 mm and breaking UK records.
- A series of depressions from the Atlantic, possibly connected to the jet stream being 'stuck'.
- High mountainous landscape combining frontal and orographic rainfall.
- Narrow bridge points and urban impermeable landscapes exacerbated flooding.
- Storm Desmond hit Cumbria in 2015.

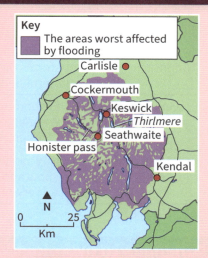

▲ **Figure 2** *Cumbria and Storm Desmond*

Social impacts
- Two people died in the floods.
- 6500 homes flooded, leading to the need for temporary shelter.
- Loss of power for 18,000 people for several days.
- Sewage mixed with water supply posed a health risk.
- Long-term impacts on mental health.

Environmental impacts
- Large amounts of flood waste from urban areas and uprooted trees choked rivers.
- River banks were eroded and ecosystems disrupted.
- Millions of tons of sediment were deposited on floodplains and in settlements downstream.

Economic impacts
- 1000 businesses were flooded. Businesses like United Biscuits in Carlisle closed for weeks, putting its future in doubt.
- 600 farms were affected with the loss of crops and livestock.
- Bridges were swept away (e.g. Pooley Bridge, Ullswater) and other road and rail links cut, with 354.8 km of carriageway damaged.
- Insurance claims exceeded £6 billion and the total direct costs of flooding were £400 to £500 million.
- House prices fell in areas at increased risk.

> **REVISION TIP**
>
> Learning key facts about case studies will help you develop explanations and evaluative points in an exam.

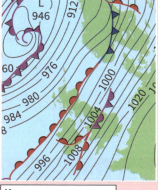

Key
- Cold fronts
- Warm fronts
- Occluded fronts

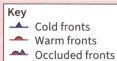

▲ **Figure 3** *A synoptic map showing the weather fronts of Storm Desmond*

Key terms Make sure you can write a definition for these key terms

channelise   depression   jet stream   monsoon   realign

 **KNOWLEDGE**

7 The water cycle and water insecurity

7.7 Climate change

Climate change and the hydrological cycle

- **Climate change** has significant impacts on the hydrological cycle locally and globally: as the planet warms up, the risks of droughts or floods increases.
- Every area may experience a different effect, with different scenarios suggested for polar and tropical regions.

Changes to hydrological inputs and outputs

Input/Output	Possible changes	Trend
Precipitation (input)	Increased evaporation resulting in more frequent and intense storms (heavy precipitation). Reduced precipitation away from storm belts and during the summer in mid-latitude areas.	Mixed
Evapotranspiration (output)	Evaporation will increase leading to more atmospheric moisture. Warmer temperatures and increased carbon dioxide encourage plant growth and higher transpiration.	Significant increase
Channel runoff (output)	May increase or decrease depending upon the precipitation. Snow and glacial melt may rapidly increase runoff in (warming) colder areas.	Unclear

Changes to hydrological stores

Store	Possible changes	Trend
Snowpacks and glaciers	A reduction in the total amount of snow mass and reduced snow-forming season. Significant glacial retreat.	Decrease
Reservoirs and lakes	May decrease in warmer climates due to increased evaporation but heavily dependent on human activity.	Unclear
Permafrost	Increased reduction through thawing, especially in northern latitudes, affecting ground water and channel flow.	Decrease
Soil moisture	Increased drying out and risk of desertification in warmer areas.	Decrease

Changes to hydrological flow

Flow	Possible changes	Trend
Channel and stream flow	An increase in areas experiencing more precipitation and a decrease in drier areas.	Mixed
Groundwater flow	Heavily dependent on human activity.	Unclear

Key terms Make sure you can write a definition for these key terms

climate change   El Niño   La Niña   permafrost

The role of ENSO and global warming on the system

- Oscillations such as the El Niño Southern Oscillation (ENSO) can change a region's climate in the short term.

- ENSO events may become more extreme and switch more rapidly, meaning droughts will follow floods and floods occur after droughts.

- The overall level of climate uncertainty will increase and feedback cycles may lead to greater imbalance within the system.

REVISION TIP

If ENSO is hard to understand, prioritise retrieving the impacts on climate.

El Niño Southern Oscillation (ENSO)

- ENSO is the short-term climate oscillation between **El Niño** and **La Niña** in the Pacific region.

- Normally, trade winds move ocean currents west, resulting in warmer waters around the Western Pacific at the end of summer (bringing heavy rain) and cooler currents around South American (leading to dry, stable conditions).

- La Niña is an intensified version of normal conditions, with very heavy rain around the Western Pacific and drought in South America.

- El Niño reverses this with trade winds moving ocean currents east, leading to unusually heavy rain in South America and dry conditions around the Western Pacific.

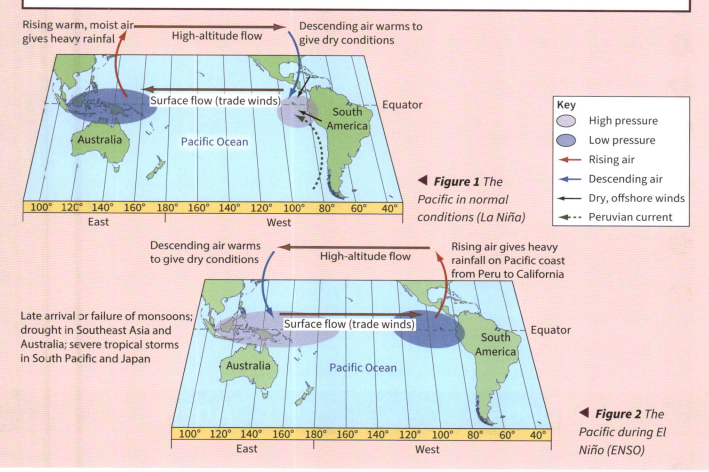

Figure 1 The Pacific in normal conditions (La Niña)

Figure 2 The Pacific during El Niño (ENSO)

Projections of future drought and flood risk

- Flood risk is not expected to increase but specific events may be 10% more intense as rain falls onto ground made less permeable by climate change.

- Droughts will likely become longer, more severe, and affect more areas, possibly exacerbated by human activity.

- However, future changes to climate are uncertain due to:
  - The complexity of factors such as ENSO, future human activity, and exposure to flood risk areas.
  - An incomplete understanding of Earth's systems, including the nature of positive and negative feedback and the role of different tipping points.
  - An incomplete data record available to analyse flood and drought risk trends.

⇄ RETRIEVAL

Learn the answers to the questions below, then cover the answers with a piece of paper and write as many as you can. Check and repeat.

Questions | Answers

#	Question	Answer
1	What type of drought relates to a deficit in precipitation?	Meteorological
2	Define the term 'tipping point'.	A threshold at which a system will irreversibly change and reach a new stable but altered state
3	What term describes trees shedding leaves during drought?	Forest stress
4	Give two physical causes of floods.	Two from: depressions / intense storms / monsoons / snowmelt
5	What is river realignment?	When original meanders on a river are straightened
6	What is affected if a river is channelised?	Flow speeds up and flood risk may increase downstream
7	How might climate change affect permafrost?	Decrease due to thawing
8	What short-term climate oscillation affects the Pacific?	El Niño Southern Oscillation (ENSO)
9	What conditions does La Niña bring to South America?	Drought (and cooler oceans)
10	Give one reason why future projections of drought and flood risk are uncertain.	One from: complexity of factors / incomplete understanding of systems including feedback cycles / incomplete data record

Put paper here

Previous questions

Now go back and use these questions to check your knowledge of previous topics.

Questions | Answers

#	Question	Answer
1	Under what conditions can water be stored in an aquifer?	If the rock is permeable
2	Define the term 'recharge'.	When soil regains lost moisture because precipitation is higher than potential evapotranspiration
3	What two variables are represented on a hydrograph?	Rainfall and discharge
4	What human factors could increase overland flow?	Deforestation, changing land use, urbanisation
5	Name one physical factor affecting hydrograph shape.	One from: size of basin / shape of basin / drainage density / rock type / soil / relief / vegetation

Put paper here

⚙ KNOWLEDGE

7 The water cycle and water insecurity

7.8 Causes of water insecurity

Mismatch between water supply and demand

- **Water insecurity** occurs when areas experience a lack of freshwater due to demand exceeding supply.

Increasing demand	Decreasing supply
Population growth	Aquifer supply which may be non-renewable
Increasing standards of living in emerging countries like China	Poor governance
Heavier use for industry and irrigation	Climate change

- A lack of freshwater is an increasing global problem, with lower-income communities worst affected, creating a **water gap** between those who have reliable access to freshwater and those who don't.

- **Water stress** is not having access to water for domestic use (below 1700 m³ per person) for a period of time.

- **Water scarcity** is continuous water shortages and deficit (below 1700 m³ per person), causing major impacts on the environment, economy, and quality of life.

  — **Physical scarcity**: water supply does not meet the demand.

  — **Economic scarcity**: people cannot afford potentially available water.

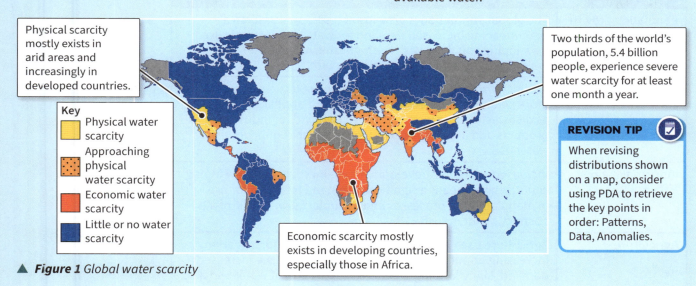

Physical scarcity mostly exists in arid areas and increasingly in developed countries.

Key
- Physical water scarcity
- Approaching physical water scarcity
- Economic water scarcity
- Little or no water scarcity

Two thirds of the world's population, 5.4 billion people, experience severe water scarcity for at least one month a year.

REVISION TIP 📝

When revising distributions shown on a map, consider using PDA to retrieve the key points in order: Patterns, Data, Anomalies.

Economic scarcity mostly exists in developing countries, especially those in Africa.

▲ *Figure 1 Global water scarcity*

Physical causes of reduced water supply

Climate variability	Physical location	Salt water encroachment at coast
Precipitation patterns determine available water supply. High-pressure zones see low rain and snowfall, although this may vary seasonally and in longer-term cycles, affected by oscillations. Global warming may reduce precipitation in some areas.	Areas far inland will generally see low precipitation. Some areas, such as the Death Valley, USA, may experience a rain shadow and be extremely dry. The geology of an area can also determine the extent of rivers and lakes and access to groundwater aquifers.	Seawater can be drawn into aquifers under the ground when there is a deficit of freshwater, possibly due to over-abstraction. This reduces the available freshwater resource.

⚙ KNOWLEDGE

7 The water cycle and water insecurity

7.8 Causes of water insecurity

Human causes of reduced water supply

Over-abstraction	Agricultural contamination	Industrial pollution	Sewage
Unsustainable abstraction for domestic use, industry, and irrigation can lead to rivers and lakes drying up. Aquifers are also being severely depleted, leading to the water table dropping.	Toxic pesticides and chemical fertilisers used on farmland contaminate groundwater supply. **Eutrophication** can occur when nitrates enter lakes and rivers, leading to the pollution of large areas.	Heavy metals, such as cadmium, from mining or factories may end up in surface and groundwater, making it unfit for human consumption.	Raw sewage contaminates water supply and increases the risk of waterborne diseases. This is a major issue for water-insecure countries, with 840,000 people dying each year.

Increasing demands on water supply

Demand is dramatically increasing in many areas and is expected to outstrip the supply of freshwater by 40% by 2030.

Demand for water is increasing due to:	
Population growth and urbanisation	The global population is growing at 0.9%, or 80 million a year, meaning increased demand for water for domestic use. Developing countries with the highest levels of water scarcity are seeing the highest population growth.
Improving living standards	The increasing middle class in emerging countries demands more water for direct use in the home (for appliances such as dishwashers), as well as through demand for industrial products and a wider variety of foods which require intensive water use.
Industrialisation	Increased economic development in emerging and developing countries has meant a significant demand for water for use in manufacturing processes and as coolants, with 40% of all abstraction for industry.
Agriculture	Agriculture uses 70% of the world's freshwater for irrigation. Increasing populations and demand for meat have increased demand. Many farming practices are wasteful of water and also pollute supply.

Key terms Make sure you can write a definition for these key terms

economic scarcity   eutrophication   physical scarcity
salt water encroachment   water gap   water insecurity
water scarcity   water stress

7.9 Consequences of water insecurity

Price of water

- Over 1.6 billion people face economic water scarcity: they cannot afford access to clean water, also known as **potable water**.
- Where the price of water is too high, there is **water poverty**.

The price of water in a country will vary due to:

- Costs of obtaining water: water from aquifers or distant reservoirs is more expensive to access.
- Lack of infrastructure: rapidly growing informal settlements may exceed the capacity to keep up with demand for piped water. Water may only be accessible by tanker or in bottles.
- **Privatised** water companies: private services charge fees to make profit, and overcharging may occur if a company has a **monopoly** or it is badly governed.
- Government policies: some governments may charge more to encourage efficiency or to reinvest in ecosystem health.

Bolivia: a high water price

- The water system in Cochabamba, Bolivia, was privatised in 1999.
- This led to a 35% increase in prices and increased water poverty.
- There were widespread protests against the new provider, Aguas del Tunari.

Water poverty index (WPI)

The **water poverty index (WPI)** can be used to measure the degree to which countries are successfully utilising their water supply.

- It is based on five components, each scoring up to 20:
  1. Resources: the physical availability of water resources.
  2. Access: the time and distance needed to obtain safe water.
  3. Capacity: how well the water is managed.
  4. Use: how economically water is used.
  5. Environment: the ability to sustain ecosystems.
- Values can be plotted, and an overall score generated, for a country or area.
- Countries at different development levels can be compared to monitor progress.

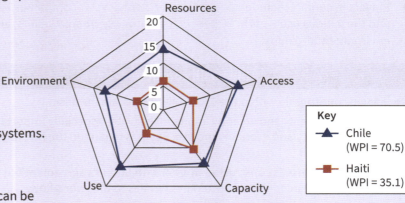

▲ *Figure 1 WPI showing Haiti and Chile*

Importance of water supply for economic development

Industry	Energy supply	Agriculture
• Industries, especially chemical and electronic, require water for manufacturing, cleaning, and cooling. Developed countries use 59% of their total water this way.	• All thermal power stations (supplying 80% of electricity used globally) require vast amounts of water; hydroelectric plants run on water alone.	• High crop yields are vital for food security and economic prosperity. These require abstraction of water for irrigation.
• Without cheap or accessible water, many manufacturing industries couldn't function and would close.	• A reduction in supply could mean short-term blackouts and long-term closure of heavy industries.	• Food shortages quickly arise with lack of water, leading to famine and civil unrest.

⚙ KNOWLEDGE

7   The water cycle and water insecurity

7.9 Consequences of water insecurity

Importance of water supply for human wellbeing

<table>
<tr><th colspan="1">Sanitation</th><th>Health and food preparation</th></tr>
<tr><td>

- Clean water is required for hygiene, washing, cleaning, and ultimately a good quality of life.
- Poor **sanitation** can lead to debilitating diseases, such as cholera.

</td><td>

- Clean hands, and food washed in clean water, prevents the spread of disease.
- **Waterborne diseases** affect life expectancy, education, and the ability to work productively leading to increased poverty.
- Stagnant water can also be a home to vectors for parasitic diseases such as malaria.

</td></tr>
</table>

Potential for conflicts

- Water insecurity can lead to **conflicts** within a country and between countries if a drainage basin crosses international borders (**transboundary** conflicts).
- The 'fair share' of water a country deserves is controversial, especially if a country constructs dams which limit the downstream flow to other countries.
- Players in conflict include national governments, water companies, and the various users of the water such as farmers, urban dwellers, and environmental groups.

REVISION TIP

Prioritise the retrieval of the conflict elements in examples. Remember that conflict doesn't always mean war.

Nile basin conflicts

- Eleven countries lie within the 6700-mile-long drainage basin.
- Seasonal droughts and an arid climate make water especially important for food security and economy growth. Egypt is reliant on the Nile for 93% of its water.
- The Aswan Dam was constructed in 1959 with Sudan's agreement but not that of other countries affected by changes to the river's flow.
- All countries except Egypt and Sudan agreed to share water fairly in 2010.
- Ethiopia began building the Grand Renaissance Dam on the Blue Nile tributary, from where 85% of the Nile's waters flow, in 2011.
- Egypt is alarmed at the possible loss of water resources (25%) and impact upon power generation at Aswan when the **mega dam** is filled, and have hinted at armed conflict.

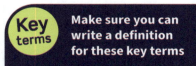

Key terms Make sure you can write a definition for these key terms

conflict   mega dam   monopoly   potable water   privatise   sanitation   transboundary   waterborne diseases   water poverty   water poverty index (WPI)

7.10 Managing water insecurity

Approaches to managing water

- **Hard engineering** provides a **techno-fix** solution to water insecurity by building solid structures to control rivers. These schemes are usually expensive, require high maintenance, and are not always sustainable.
- **Sustainable schemes** aim to ensure resources are protected for future generations by conserving supplies and reducing demand, as well as working more with natural processes.
- Water sharing **treaties** and agreements based on entire drainage basins may incorporate the methods above but need political engagement and good governance to work.

Hard engineering schemes

Water transfers

The diversion of water from an area of surplus to deficit, sometimes across huge distances, using a series of canals/canalised rivers, pumps, and pipelines.

Pros	Cons
☑ Provides much needed water supply	☒ Reduces water availability in source area
☑ Reduces costs long-term	☒ May submerge huge areas
☑ Reduces the need for abstraction from aquifers	☒ Expensive infrastructure projects
	☒ Relies upon current flow data

The South–North Water Transfer Project, China
The project aims to move 45 billion cubic metres of water a year from the south to the more arid north of China and the capital, Beijing. However:

- The transfer of water could lead to shortages in the south.
- The costs of US$62 billion have been criticised and corruption may be an issue.
- Over 1.5 million people have been resettled from submerged land.

Mega dams

The construction of large dams to control water flow and to create reservoirs for storage.

Pros	Cons
☑ Creates a vast store of water for domestic, industrial, and agricultural use	☒ Expensive infrastructure projects
☑ Controls river flooding	☒ Huge areas of land need to be flooded for reservoirs
☑ May provide hydroelectric power	☒ Reservoirs may silt up and become contaminated

Desalination

Seawater is turned into freshwater at coastal locations using salt-separating membranes and the process of reverse osmosis.

Pros	Cons
☑ Produces potable water in areas which may lack rivers or aquifers	☒ Very expensive, with high energy demand
☑ Reliable and predictable supply	☒ Produces concentrated salt water which can affect ecosystems

⚙ KNOWLEDGE

7 The water cycle and water insecurity

7.10 Managing water insecurity

Sustainable schemes

River restoration

Restoring lakes and rivers to a more natural condition reduces the need for expensive hard engineering and maintenance, and increases water stores whilst improving ecosystems. Restoration schemes involve restoring natural meanders, reflooding areas to recreate wetlands, and reducing abstraction from aquifers.

Water conservation

Conserving water reduces demand and makes water use more efficient. **Conservation** schemes involve:

- Smart irrigation methods:
  - Drip irrigation rather than using sprinklers, saving water and money.
  - Hydroponics involves the growing of crops without soil in greenhouses and drip-feeding water and nutrients.
- Recycling water:
  - The use of **grey water** from cities for agriculture reduces demand.
  - Rainwater and sewage can be purified via filtration and used as drinking water.
- Rainwater harvesting: using catchments on roofs and drains to store rainwater.

Integrated drainage basin management

Large rivers can also be managed in a holistic manner through **integrated water resource management (IWRM)**. This aims to:

- coordinate the development and management of resources between the different players involved
- consider the whole drainage basin and natural hydrology
- maximise economic and social services in an equitable manner
- protect the sustainability of ecosystems.

IWRM of the Colorado River

- Original agreements shared water unfairly, leading to Mexico only receiving 10% of the basin's water and the State of California taking 20% more than its allocation.
- IWRM has led to greater agreement between stakeholders including indigenous American peoples who were not receiving a fair share. A 2012 agreement gives Mexico the right to some water stored in Lake Mead in the USA.
- A new agreement in 2007 used current climate data to share surpluses and deficits more fairly.
- There is, however, still a long way to go.

Water-sharing treaties and frameworks

These aim to reduce the risks of conflicts and increase the efficiency of water usage.
They are governed by IGOs and have had mixed success.

Helsinki Rules

- International guidance for transboundary rivers and aquifers.
- Focused on equitable use and equitable shares of water based on holistic criteria.

UNECE water convention

- UNECE is the United Nations Economic Commission for Europe.
- Based on the IWRM principles.
- Requires players to use transboundary water in a reasonable and fair way.

EU WFD

- WFD is the Water Framework Directive.
- Focuses upon surface and groundwater health.
- Set standards for tackling pollution and supporting wildlife in the EU.

Key terms Make sure you can write a definition for these key terms

conservation   desalination   grey water   hard engineering
integrated water resource management (IWRM)   restoration
sustainable schemes   techno-fix   treaties   water transfers

RETRIEVAL

Learn the answers to the questions below, then cover the answers with a piece of paper and write as many as you can. Check and repeat.

Questions

		Answers
1	What are the two elements of water scarcity?	Physical scarcity and economic scarcity
2	What is salt water encroachment?	When seawater is drawn into freshwater aquifers
3	Give one human cause of reduced supply which could lead to water insecurity.	One from: abstraction / agricultural contamination / industrial pollution / sewage
4	Define the term 'potable water'.	Clean water (drinkable)
5	How many components make up the WPI?	Five
6	Define the term 'hard engineering'.	The building of solid structures (to control rivers)
7	Give the two benefits of desalination.	Reliable; produces water in areas without accessible water
8	Give one method of water conservation.	One from: smart irrigation (drip or hydroponics) / recycling water / rainwater harvesting
9	Give one aim of integrated water resource management (IWRM).	One from: coordinate the development and management of resources between the different players involved / consider the whole drainage basin and natural hydrology / maximise economic and social services in an equitable manner / protect the sustainability of ecosystems
10	Name one water-sharing treaty or framework.	One from: Helsinki rules / UNECE water convention / EU WFD

Put paper here

Previous questions

Now go back and use these questions to check your knowledge of previous topics.

Questions

		Answers
1	What is river realignment?	When original meanders on a river are straightened
2	Define the term 'tipping point'.	A threshold at which a system will irreversibly change and reach a new stable but altered state
3	If inputs are higher than outputs, the water budget is in surplus. True or false?	True
4	What conditions does La Niña bring to South America?	Drought (and cooler oceans)
5	Give two physical causes of floods.	Two from: depressions / intense storms / monsoons / snowmelt

Put paper here

Exam-style questions

1   Study **Figure 1**. Suggest **one** possible consequence on the hydrological cycle of the land use change shown in Figure 1.

(3)

EXAM TIP

This question is 3 marks, which means giving the one consequence and then two further developed points. Divide your answer into three sentences if that helps.

▲ **Figure 1** *Deforestation in Canada*

2   Study **Figure 2**. Explain the relationship between potential evapotranspiration and soil moisture in October.   **(3)**

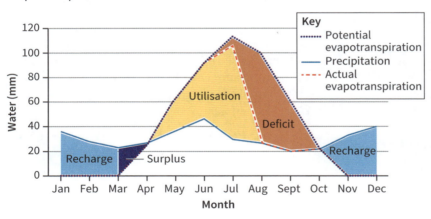

▲ **Figure 2** *A water budget from South Okanagan, Canada*

3   Study **Figure 3**. Explain **one** possible reason why the water management strategy shown in Figure 5 may be controversial.

(3)

EXAM TIP

Sometimes examiners provide resource examples that you may not have studied. This does not matter as you just need to use the resource as a stimulus to understand the type of strategy and maybe apply your own example.

◄ **Figure 3** *A satellite image of the Grand Ethiopian Renaissance Dam under construction*

4 **Figure 4** shows a drought index for the USA. Explain **one** reason for the trends shown. **(3)**

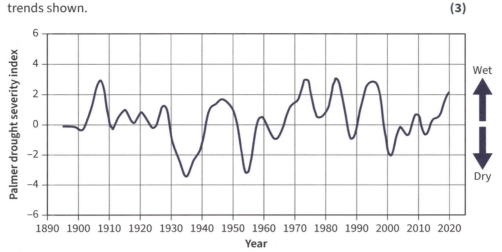

▲ **Figure 4** *A drought index for the USA*

5 Explain how human action may exacerbate flood risk. **(6)**

6 Explain why a drainage basin can be considered an open system. **(6)**

7 Explain why the future projection of droughts and floods may be uncertain. **(6)**

8 Explain the importance of water supply for economic development and human wellbeing. **(6)**

9 Explain why the price of water varies between different areas of the world. **(8)**

10 Explain the physical factors which can lead to increased water insecurity. **(8)**

11 Explain how water management can be made more sustainable. **(8)**

12 Explain why droughts may be becoming an increasing problem for people and the environment. **(8)**

> **EXAM TIP** 🎯
>
> There are numerous factors to consider here and the examiner is ideally looking for three explained ideas.

> **EXAM TIP** 🎯
>
> The 8-mark explain questions are just extended 6-mark questions. Look to add another idea and remember you don't have to conclude.

13 Study **Figures 5** and **6**, which show two river catchments and hydrographs. The two catchments have been affected by the same storm at the same time.

Assess the likely reasons for the contrasting shape of the storm hydrographs. **(12)**

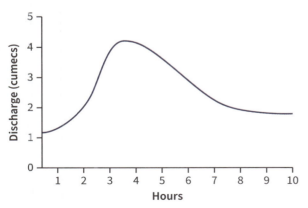

▲ **Figure 5** *Catchment A*

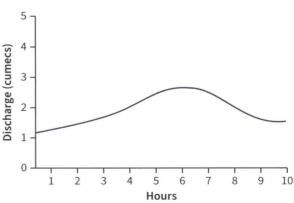

▲ **Figure 6** *Catchment B*

14   Assess the impact of climate change on stores within the
hydrological cycle.   **(12)**

15   Assess the physical and human causes of water insecurity.   **(12)**

16   Study **Figure 7**, a map showing projected water stress in 2040.

Assess the role of human factors in influencing the pattern of future
water stress.   **(12)**

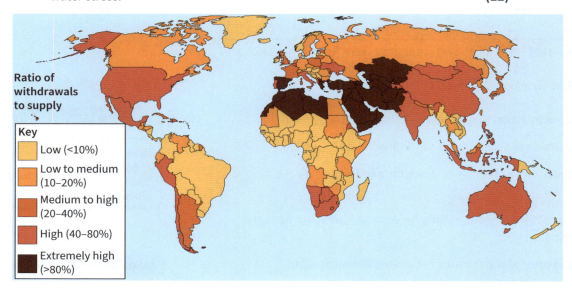

17   Evaluate the view that conflicts between most transboundary
water users are likely to increase in the future.   **(20)**

EXAM TIP

Always make sure you
give a balanced answer,
featuring both sides of
the argument. Each point
you make needs to be
evaluated and contributes
to a conclusion, where
a clear decision is made
and justified.

18   Evaluate the view that the management of water supply is
increasingly unsustainable.   **(20)**

19   Evaluate the risks and consequences of an inadequate water supply
for people and the environment.   **(20)**

EXAM TIP

This question requires
several components
to be covered: risks
and consequences of
inadequate water as well
as a focus on both people
and the environment. All
should feature as part
arguments in your answer.

20   Evaluate the view that hard engineering projects are always the
most successful strategies when it comes to solving water insecurity
for people and the environment.   **(20)**

21   Evaluate the view that changes to the water cycle pose a greater risk to
humanity than changes to the carbon cycle.   **(20)**

EXAM TIP

Sometimes the 20-mark
'Evaluate' question is
a mixture of the water
and carbon topics. If
this happens, then look
to balance your answer
between the two.

⚙ KNOWLEDGE

8 Carbon cycle and energy security

8.1 The carbon cycle

Carbon stores

- Carbon is stored in solids, liquids and gases.
- Most global carbon is locked in terrestrial stores as part of the long-term geological cycle (slow carbon cycle).
- It occurs in biotic (living) items, e.g. trees and humans, and abiotic (non-living) things, e.g. sedimentary rock.

- Carbon stores vary in size and are measured in petagrams (Pg) (1Pg = 1 gigaton).
- Carbon is transferred (**fluxed**) between stores by processes such as rock weathering, volcanic eruptions and photosynthesis; these fluxes are measured in Pg/year.
- These occur on short and long timescales.

Carbon store	% of global total	Store type	Changes and examples
Biosphere (living organisms)	0.0012%	As a solid in trees, plants, and animals	Areas of high net primary productivity, e.g. tropical and temperate forests (Amazon and Adirondack).
Atmosphere (air)	0.0017%	As a gas – carbon dioxide (CO_2)	This store is increasing due to burning of fossil fuels which transfers carbon from the lithosphere to the atmosphere.
Pedosphere (soil)	0.0031%	As a solid	Found in soils which contain organic matter, e.g. peat is 60% carbon.
Hydrosphere (water)	0.038%	In a liquid	90% of the carbon stored in oceans is in the form of hydrogen carbonate ions.
Lithosphere (rocks and minerals)	99.906%	As a solid, liquid and gas	Sedimentary rocks, e.g. limestone or coal, contain remains of organisms such as their skeletons and shells. Coal is up to 80% carbon, oil up to 85% and gas 75%.

Key terms Make sure you can write a definition for these key terms

atmosphere   biosphere   chemical weathering   flux   hydrosphere
lithosphere   pedosphere   volcanic outgassing

8 Carbon cycle and energy security

8.1 The carbon cycle

The geological carbon cycle

Two sets of processes lead to a large store of carbon in the lithosphere.

Formation of shales and fossil fuels

1. On the ocean floor, fine grained sediments and dead organisms (algae, phytoplankton) accumulate and degrade.

⬇

2. Accumulation continues with newer sediments and organisms on top pressurising and heating up older sediment at the bottom.

⬇

3. Compression continues as more sediment and organisms sink to the ocean floor. This leads to chemicals reactions and the formation of insoluble geopolymers.

⬇

4. At depths of more than 1000 m, a thermocatalytic process creates source rock rich in organic matter.

⬇

5. At depths greater than 2500 m, crude oil and natural gas can form.

Whilst the lithosphere is the largest carbon store, the processes which transfer carbon are extremely slow. It takes 100–200 million years to move carbon between rocks, soil, ocean, and atmosphere. On average, 0.01 to 0.1 Gt of carbon move through the slow carbon cycle every year.

Sedimentary carbonate rock formation

1. On the ocean floor, fine grained sediments mixed with shells and skeletons settle. This calcareous ooze is rich in calcite.

⬇

2. Continued accumulation compresses and compacts this ooze, and it cements together.

⬇

3. When the compacted ooze reaches temperatures over 40°C, it goes through a process of diagenesis and lithification to form sedimentary rock.

⬇

4. The dead organisms are the main carbon source in this rock.

REVISION TIP

These processes are sequential. If you can write physical processes like this sequentially, you will improve your chances of getting good marks.

REVISION TIP

This is important in understanding why burning large amounts of fossil fuels leads to a long-term change in the carbon cycle balance.

Geological fluxes of carbon (atmosphere to lithosphere to atmosphere)

There are two main processes: **chemical weathering** of rock and **volcanic outgassing**.

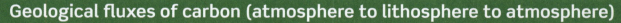

6. Some of this CO_2 rises back to the surface within the heated magma and then is degassed through volcanoes, returning CO_2 to the atmosphere.

1. Water reacts with atmospheric CO_2 to form carbonic acid.

CO_2 in the atmosphere

5. Some limestone and carbonate sediments will be subducted under continental crust at convergent plate margins where CO_2 is released by heat.

4. Calcium carbonate is also precipitated by marine organisms. Carbonate sediments are formed.

2. When this reaches the surface as rain it reacts with minerals in surface rocks and dissolves them into component ions.

3. Calcium ions are transported by rivers from land to oceans combine with carbon ions to form calcium carbonate. Deposition and burial turns calcite sediment into limestone.

▲ *Figure 1 The geological carbon cycle interacts with the rock cycle*

8.2 Carbon sequestration

The biological carbon cycle

- Marine and terrestrial processes transfer carbon from the atmosphere to the hydrosphere, biosphere and pedosphere which are carbon sinks. This happens on shorter timescales (fast carbon cycle).

- Photosynthesis is responsible for the largest transfer of carbon. Terrestrial primary producers transfer 123 PgC/yr and marine phytoplankton 80 PgC/yr.

- Respiration returns 197.1 PgC/yr from land and sea to the atmosphere.

Ocean carbon fluxes

Marine carbonate pump

Atmospheric carbon is **sequestered** through **photosynthesis** by photoplankton in surface waters.

↓

CO_2 also dissolves in the ocean and reacts chemically with water to form solid calcium carbonate in the shells and skeletons of marine organisms: e.g. crabs, zooplankton and oysters.

↓

These organisms sink to the deep ocean when they die and accumulate on the sea bed forming calcareous ooze and limestone.

↓

Respiration and decomposition return some gaseous carbon to the atmosphere.

> **REVISION TIP**
>
> To help you revise physical geographical processes, create a numbered list which is chronological and in which steps are connected to each other.

> **REVISION TIP**
>
> Note the similarities and differences between the processes of terrestrial and marine carbon transfers.

Thermohaline circulation

◄ **Figure 1**
Thermohaline circulation

1. Dissolved carbon is transferred from the tropics to the poles and from surface to deep ocean through the thermohaline circulation.

2. This movement is caused by differences in water density due to temperature and **salinity** (lower temperatures and higher salinity increase water's density).

3. Colder water at higher latitudes and deep water under pressure can hold more dissolved CO_2.

4. Therefore, oceans such as the Southern Ocean are important carbon sinks (holding up to 25% of carbon sequestered from the atmosphere) whereas warmer oceans such as the tropical Atlantic are carbon sources.

8 Carbon cycle and energy security

8.2 Carbon sequestration

Terrestrial carbon fluxes

Plants

Plants use solar energy to covert CO_2 and water into carbohydrates which lead to growth.

This process is greatest in the tropics where limiting factors such as changes in daylength, lower temperatures or lack of water are less significant. Therefore, **net primary productivity (NPP)** is high. Conversely, NPP is low at the poles.

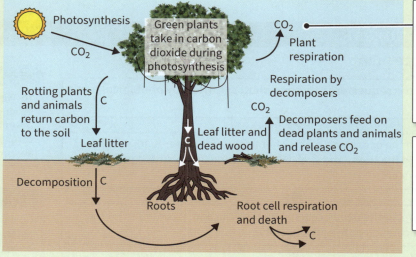

Photosynthesis

CO_2

Green plants take in carbon dioxide during photosynthesis

Rotting plants and animals return carbon to the soil

C

Leaf litter

Decomposition C

Roots

CO_2

Plant respiration

Respiration by decomposers

CO_2

Leaf litter and dead wood

Decomposers feed on dead plants and animals and release CO_2

Root cell respiration and death

C

In **respiration**, organisms convert carbohydrates into energy water and CO_2. The CO_2, water and some of the energy are released to the atmosphere.

However, about 1000 times more CO_2 is taken from the atmosphere than is released back to it.

▲ *Figure 2 The tree carbon cycle*

Soils

Soils contain organic and inorganic carbon.

- Carbon is transferred into the soil in the form of organic compounds released into the soil by plant roots or through the decay of plant material or organisms when they die.

- Microbial breakdown of the organic matter releases the nutrients which plants use to grow.

- During this process of decomposition, some carbon is released as carbon dioxide through soil respiration, while other carbon is converted into stable organic compounds that are locked into the ground.

- Soils with a high percentage of organic matter, e.g. peat, are major carbon sinks.

The importance of mangrove forests

An estimated 1.5 metric tons of carbon per hectare per year is sequestered in biomass through photosynthesis.

Mangrove forests are found in tropical and sub-tropical tidal coastal areas where NPP is high.

Respiration releases some CO_2 back into the atmosphere.

Large amounts of organic carbon are stored in the soil. Little respiration occurs as the soil is anaerobic (lacking oxygen) and only 10% of it is organic matter.

Key terms

Make sure you can write a definition for these key terms

carbonate pump   net primary productivity   photosynthesis
respiration   salinity   sequester   thermohaline circulation

8.3 Balancing the carbon cycle

The natural and enhanced greenhouse effects

- The **natural greenhouse effect** means that life can exist on planet Earth; without it, Earth's average temperature would be −6 °C rather than +15 °C.

- When **solar energy** reaches the atmosphere, some is reflected by clouds and (due to its short wavelength) about half of it passes through gases in the lower atmosphere to reach the Earth's surface.

- As the surface warms, it radiates longer-wavelength radiation (heat) into the atmosphere. This is absorbed by **greenhouse gases** such as water vapour, CO_2 and methane.

- Absorption leads to warming of the atmosphere.

- Increasing amounts of greenhouse gases, primarily from the burning of fossil fuels, increases the amount of absorbed heat, leading to the **enhanced greenhouse effect**.

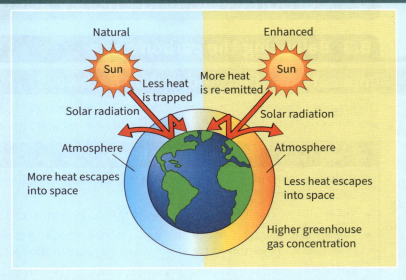

▲ **Figure 1** The natural and enhanced greenhouse effects

REVISION TIP

Make sure you know the difference between the natural and enhanced greenhouse effects.

Greenhouse gases

Greenhouse gas	Anthropogenic sources	Global warming potential (relative to CO_2)	% of total emissions
Carbon dioxide (CO_2)	Burning fossil fuels (coal, natural gas, and oil), solid waste, deforestation and soil degradation, chemical reactions (e.g. cement production)	1	76
Methane (CH_4)	Production and transport of fossil fuels, livestock and rice farming, decay of organic waste in municipal landfills	21	16
Nitrous oxide (N_2O)	Combustion of fossil fuels, chemical fertilisers, treatment of wastewater	310	6
Fluorinated gases e.g. Hydrofluorocarbons	A variety of household, commercial, and industrial applications and processes	100s to 10,000s	2

Implications of fossil fuel consumption

- Pre-Industrial Revolution fast fluxes of carbon were fairly constant and the cycle was balanced with CO_2 concentrations around 180–290 ppm (parts per million).

- Post-1750 anthropogenic (human) activity has transferred large amounts of carbon from fossil stores – long-term in nature – to short-term stores and fluxes.

- CO_2 concentrations in the atmosphere are now around 420 ppm.

- Excess carbon is stored in the atmosphere (leading to warming), the oceans (leading to acidification) and **terrestrial ecosystems** (which can lead to increased productivity due to better soil health).

- Current warming is about 8 times faster than the warming after the last ice age.

8.3 Balancing the carbon cycle

Implications of altered carbon pathways

The process of fossil fuel combustion has changed the balance of carbon pathways and stores on Earth with implications for the climate, ecosystems and the hydrological cycle.

Climate

- High probability of increasing droughts in the Sahel, the Mediterranean, and Southern Africa.
- Thermohaline circulation may weaken in the North Atlantic and Southern Ocean.
- The world is now 1.2 °C hotter than it was in the second half of the 19th century. More heat and moisture in the atmosphere may lead to stronger storm activity.
- Extreme summer temperatures in Europe, China and elsewhere contributed to 2022 being the fifth-hottest year on record. Global average temperatures have been increasing since the 1980s.
- The average Arctic temperature has increased at twice the global average over the last 200 years leading to the extent of sea ice decreasing by 12.2% per decade.

Ecosystems

- Rates of extinction could rise 15–40% due to habitat changes and limited adaptability of species. Especially so in polar regions where amplified climate changes could impact e.g. polar bears, caribou, and emperor penguins.
- Changes in biodiversity as habitats shift poleward, into deeper ocean waters, or to higher altitudes as climates warm. For example, by 2080, shifting temperatures and costal flooding may reduce the range of the Pacific Golden Plover in North America.
- In the ocean, more frequent heatwaves and long-term trends of acidification increase stress on many organisms and ecosystems and may lead to mass mortality. Up to 80% of coral reefs may be bleached and die.
- Tundra biome and taiga forests rich in food sources such as blueberries, sedges and lichen for grizzly bears are threatened as permafrost thaws and slumping expands, transforming parts of the landscape into mud, silt, and peat.

Hydrological cycle

- Where precipitation is reduced and evaporation increases, rivers such as the River Po in Italy will dry up.
- Small alpine glaciers such as that on Mount Kilimanjaro in Tanzania will ablate and disappear.
- Extreme heavy precipitation events will become common over northern hemisphere land areas, e.g. rainfall events in the UK exceeding 20 mm/hr could be four times as frequent as in the 1980s by 2080.
- A shift of sub-tropical high pressure areas northwards may cause a 20–23% decrease in water availability in the Mediterranean.

 Key terms | **Make sure you can write a definition for these key terms** | enhanced greenhouse effect   greenhouse gas
hydrological cycle   natural greenhouse effect
solar energy   terrestrial ecosystem

 RETRIEVAL

Learn the answers to the questions below, then cover the answers column with a piece of paper and write as many as you can. Check and repeat.

	Questions		Answers
1	List the five carbon stores in order of size with the largest first.	Put paper here	Lithosphere, hydrosphere, pedosphere, atmosphere, biosphere
2	Which type of rock is carbon found in?		Sedimentary
3	Which process transfers the most carbon between stores?		Photosynthesis
4	Which fluxes of carbon are geological?	Put paper here	Chemical weathering and volcanic outgassing
5	What unit is used to measure the mass of carbon in a store?		Petagrams
6	On what time scale does the geological carbon cycle operate?	Put paper here	Long term
7	What are the two main processes which transfer carbon from the atmosphere to the lithosphere and back to the atmosphere?		Volcanic outgassing and chemical weathering of rocks
8	Which organisms sequester carbon dioxide into the ocean through photosynthesis?	Put paper here	Phytoplankton
9	What pump transfers carbon from the surface to the deep ocean?		Carbonate
10	What is the name of the circulation that transfers dissolved carbon around the oceans?	Put paper here	Thermohaline
11	Differences in temperature and salinity cause a difference of what in the ocean?		Density
12	Which process transfers the most carbon between stores?	Put paper here	Photosynthesis
13	Does terrestrial radiation have a short or long wavelength?		Long
14	Name four greenhouse gases.		Carbon dioxide, methane, nitrous oxide, fluorinated gases
15	Which greenhouse gas has the most warming potential?	Put paper here	Fluorinated gases
16	What is the approximate current concentration of carbon dioxide in the atmosphere?		420 ppm

8   Carbon cycle and energy security

8.4 Energy security

Energy security

| reliable, uninterrupted supplies | | more affordable energy for consumers |

A country that is energy secure has

| abundant domestic primary sources | | imported sources from allies |

Importance of energy security

- It supports industrial activity, powers businesses, and drives economic growth and development.
- It can keep energy prices low, enabling businesses to thrive and consumers to spend less disposable income on energy bills.
- Affordable energy supports social stability.

Energy consumption and sources

Energy consumption is usually expressed in per capita terms using one of the following units:

- kilograms of oil equivalent per year (kgoe/yr)
- gigajoules per year (GJ/yr)
- megawatt hours per year (MWh/yr).

Energy intensity is a measure of how efficiently a country is using its energy, and shows how much energy is consumed per unit of GDP.

- Emerging economies have higher energy intensities as infrastructure development demands higher energy.
- HICs have falling intensities as cost-cutting squeezes more value out of established markets and manufacturing.

Primary energy

Primary energy comes from sources used to generate electricity:

- non-renewable fossil fuels such as coal, oil, and natural gas
- renewable energy such as wind, geothermal, hydroelectricity and solar
- recyclable fuels such as nuclear energy, biomass, and general waste.

Energy mix

Energy mix is the combination of primary energy sources used in a country.

- Governments try to ensure a good balance between domestic energy sources and imported energy sources.
- **Energy security** increases as dependence on imported energy sources decreases.

Factors affecting consumption of energy

Stage of economic development

- A higher level of development means higher consumption per capita due to increased wealth, demand for consumer goods, demand from industry and agriculture.
- LICs have lowest consumption per capita due to lack of access to electricity in rural areas, mainly subsistence agriculture and not many energy-intensive industries.
- RICs have rapidly increasing consumption due to FDI investment from TNCs into manufacturing industries. Domestic energy demand increases due to rising affluence in urban areas.
- HICs usually have the highest energy consumption per capita although, due to outsourcing and policies around energy efficiency, this can be decreasing.

Physical availability

- Usually decreases cost, because no transport costs, e.g. in Saudi Arabia, which has large oil reserves.
- Cheaper energy means increased consumption per capita.
- If homes are not connected to a national grid less electricity will be used, e.g. in rural areas of LICs.
- Having more domestic energy sources improves energy security by reducing reliance on imported ones.

Climate

- In continental climates with sub-zero temperatures in the winter and hot summers e.g. in parts of Canada homes, businesses, cars, all need heating or cooling.

Lifestyle

- There is a strong positive correlation between rising personal wealth and higher energy consumption.

National environmental policies

- For example, in the UK, efficiency policies around insulating homes coupled with public awareness campaigns promoting more sustainable lifestyles mean that energy consumption per capita is back to rates seen in the 1970s.

Geopolitics

- Countries reliant on imported primary energy sources, particularly fossil fuels, are more vulnerable to unstable geopolitics, e.g. Germany was heavily reliant on imported gas from Russia.
- International political instability raises energy prices in global markets leading to increased cost for consumers.

Technology

- HICs have access to advanced technology so nuclear power is an option.
- Availability, access, and use of green technologies including renewable energy have significantly improved. For example: solar panels are now cheaper, can be mounted on a rotating platform so they track the sun, and black solar cell technology can generate power even on cloudier days.

KNOWLEDGE

8 Carbon cycle and energy security

8.4 Energy security

National comparisons: Norway and UK

Energy mixes

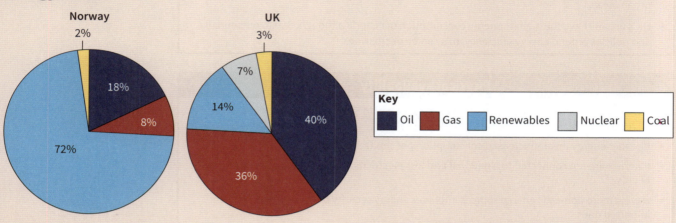

▲ *Figure 1* Norway and UK energy mixes

Similarities and differences

- Both are HICs so energy consumption is high. The UK ranked 15th and Norway 29th in the world. This difference is explained by the significantly higher population in the UK.

- Energy consumption per capita in Norway is more than double that in the UK as it has a large energy-intensive manufacturing sector, electricity is much more widely used to heat buildings and water, and the climate is colder.

- Norway has a much larger renewables sector, mostly hydroelectric as it has mountainous topography and abundant precipitation. The UK's renewable sector is growing, particularly wind, which generates about one third of electricity demand annually.

- Norway is much more energy secure than the UK as it is much less reliant on imported fossil fuels and generates more domestically sourced energy. In 2020, 87% of its energy production was exported. Most UK homes are heated by gas, about half of which is imported.

- A new nuclear power station, Hinkley Point C in Somerset, will provide low-carbon electricity for around six million homes and improve the UK's energy security.

- Both countries have ambitious targets to decarbonise their energy supplies in the future. For example, Norway has committed to cutting emissions by 55% by 2030, and achieving a net-zero economy by 2050 whilst the UK is aiming to have all electricity produced from renewable sources by 2035.

- The costs of exploitation and production change over time. The UK has large reserves of coal under the ground and an estimated 23 trillion tonnes under the sea. However, it is uneconomic to mine it at the moment. As global reserves of coal fall it may become economic to mine it again. However, both public opinion and government policies would be against this.

- Both Norway and the UK have access to deep-water drilling technology to extract oil and gas from under the North Sea.

Energy players

- An **energy pathway** is the route taken by any form of energy from its source to its point of consumption. It can involve tanker ships, pipelines and electricity transmission grids.
- Various players are involved in securing pathways and energy supplies at a range of scales.
- Each pathway has a supply end controlled by international TNC energy companies or cartels, e.g. OPEC and national governments, and a demand end controlled by governments and consumers. Along each pathway, there are companies involved in transport and processing.

TNCs	OPEC
- Examples: Gazprom, ExxonMobil, PetroChina and Royal Dutch Shell. - Nearly half of the top 20 companies are state-owned (all or in part) and, therefore, very much under government control. - Most are involved in a range of operations: exploring, extracting, transporting, refining and producing petrochemicals. - They are extremely profitable but are challenged by the recent rise of OPEC and environmental groups.	- An IGO of 15 countries which, between them, own about two-thirds of the world's oil reserves. - It can control the amount of oil and gas entering the global market and, therefore, the price of these commodities. - It aims to stabilise global oil prices whilst limiting environmental damage.
National governments	**Consumers**
- Aim to make their country energy secure through reliable supplies for their country. - Regulate the role of private companies. - Develop energy policies to, for example, reduce CO_2 emissions or improve energy efficiency.	- Include transport, industry and domestic users of energy. - Create demand with purchasing choices usually based on price. - Can choose more local, environmentally friendly options or protest and so influence awareness and policy.

> **REVISION TIP**
>
> Think of a mnemonic to help you remember the four key players in securing energy pathways and supply (TNCs, OPEC, national governments and consumers).

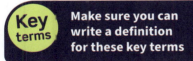

Key terms — Make sure you can write a definition for these key terms

energy consumption   energy mix   energy pathway
energy security   primary energy

 **KNOWLEDGE**

8 Carbon cycle and energy security

8.5 Fossil fuels

Fossil fuels and economic development

- Use of fossil fuels to drive economic development is still the norm.
- Fossil fuels account for over 80% of global primary energy consumption.
- Fossil fuels are not distributed evenly across the world only being found where there is favourable geology.

- There is an **energy mismatch** between where fossil fuels are found and where consumption demand is highest.
- The USA and Russia are oil and gas superpowers, while China is a coal superpower.

Energy pathways

Pipeline

- For oil and gas.
- Can be overland or on seabed.
- The East Siberia Pacific Ocean Pipeline is a **bilateral energy pathway** between Russia and China that is more than 4000 km long.

Shipping

- For oil, gas and coal
- Large shipping tankers transport more than half of global oil and gas.
- Oil spills are a risk, as are delays at **choke points**, e.g. the Suez Canal.
- Choke points are a crucial factor affecting global energy security.

Transmission lines

- Form an essential part of national grids and transport electricity from power stations via sub-stations to consumers.

Road and rail

- Mined coal is transported from mines to power stations or imported coal is shifted from the port.
- Refined crude oil in the form of petroleum is moved from the processing plant to filling stations.

Energy pathway disruption

Energy pathways are prone to disruption if:

- they are longer
- they involve **multilateral agreements** (between many countries)
- they flow through an area prone to piracy, e.g. the Straits of Malacca
- they cross tectonically active parts of the world
- they are in politically unstable regions of the world, e.g. Iraq and the Middle East
- extreme weather occurs, e.g. Hurricane Katrina destroyed oil rigs in the Gulf of Mexico.

As fossil fuel resources are depleted, the risk of disruption grows.

 Key terms Make sure you can write a definition for these key terms

bitumen   bilateral energy pathway   choke point
energy mismatch   multilateral agreements   taiga biome
unconventional fossil fuels

Geopolitics and energy pathway disruption: Russia and Ukraine

- Oil and gas have frequently been at the root of international tensions.
- Tensions between Russia and Ukraine have been high since 2004 due to political changes such as Ukraine wishing to join the EU and NATO.
- Further escalation occurred in 2014 when Russia annexed Crimea, and in 2022 following the invasion of Ukraine by Russia.
- These tensions will impact energy supplies because the EU gets a quarter of its gas from Russia, and 80% of this passes through Ukraine.

- Russia cut gas flows to the EU by around 80% between May and October 2022, leaving the EU with a significant shortfall in its energy mix.
- Energy costs have risen, as have prices of goods and services.
- Tense relationships between Russia and the EU could accelerate the move to more sustainable fuels and renewable sources.

Unconventional fossil fuels

Tar sands

Surface deposits of **bitumen** are mined. Canada is the global leader in production.

Advantages	Disadvantages
☑ Creates economic growth for rural regions. ☑ Improves Canada's energy security.	☒ Extracting bitumen is water- and energy-intensive, producing a large volume of waste. ☒ Liquid waste left in tailing ponds contains sulphates, chlorides and ammonia which may infiltrate groundwater stores and other water sources. ☒ Open mining involves removing the top layer of vegetation and carbon-rich soils to access the bitumen sands, destroying habitats in environmentally sensitive **taiga biomes**.

Fracking (shale oil and gas)

Extracts gas and oil from underground rocks such as shale. Water, chemicals and sand are pumped into the ground to break up the shale, access the hydrocarbons (shale gas or shale oil) and force them to the surface.

Advantages	Disadvantages
☑ Less polluting than coal or oil. ☑ Provided 25% of US energy needs in 2015, contributing to energy security.	☒ Wastewater needs treating due to chemical contents. ☒ May pollute groundwater aquifers. ☒ Earthquakes of low magnitude may occur.

Deep-water oil drilling

Oil reserves under the ocean floor are exploited as more accessible reserves under land and shallow oceans run out. Brazil has such offshore reserves.

Advantages	Disadvantages
☑ Using new reserves lengthens the time until peak oil (the time when the maximum rate of global oil production is reached) and keeps prices lower. ☑ Many engines and appliances are designed to operate on oil so continuing to extract oil would avoid large changes to many important engines, e.g. motor vehicles, planes.	☒ More oil reduces momentum to move to renewable energy. ☒ Extracting and burning it continues to transfer carbon from a long-term geological store into the atmosphere which contributes to global warming.

KNOWLEDGE

8 Carbon cycle and energy security

8.6 Alternatives to fossil fuels

The UK's changing energy mix

Changes	Explaining these changes
• 1970s: 91% of primary energy came from oil and coal. • 1980s: 22% came from gas; proportion of nuclear power had increased. • All coal-fired power stations are due to close in 2025 and production of oil and gas has been decreasing since 2000. • In 2022, wind was the second largest source of electricity (almost 30%) after gas. • High sunshine hours means solar contributes up to 4.4% in summer. • Currently there is no exploitation of unconventional fossil fuels.	• The discovery of oil and gas under the North Sea in the 1960s and 70s increased their use. • Recent government decisions to start drilling in the Roseback Oilfield may increase the use of oil and increase energy security. • Nuclear will increase when Hinkley Point C in Somerset opens in the next decade. This facility will provide 7% of electricity demand. • Renewables are increasing as the government aims for the country to have reached **net zero** by 2050. • As energy efficiency improves, energy consumption per capita will decrease.

Renewable and recyclable energy

These have economic, social and environmental costs and benefits.

Wind power	Wave power	Solar power
✓ Low running costs once installed. ✓ Plenty of suitable sites on and offshore. ✗ Birdlife can be affected. ✗ Impact the aesthetic of the landscape.	✓ Produces most electricity during winter when waves are larger and demand is highest. ✓ Technology and innovation are rapidly improving. ✗ Very expensive and no 'perfect' solution yet. ✗ Needs to survive storms.	✓ Costs are decreasing rapidly. ✓ Large potential in desert areas. ✓ Rapidly developing new technologies e.g. solar windows. ✗ Not yet very efficient. ✗ Effectiveness dependent on climate.

Can renewables replace fossil fuels?

Yes	No
✓ Could help decouple fossil fuels from economic growth if efficiency improves. ✓ Renewable energy is likely to be an important component of the future energy mix as it has a low **carbon footprint** and the technology is always improving. ✓ Each renewable resource has advantages and disadvantages, though the disadvantages will decrease as the technologies are improved. ✓ If investment continues, there are endless possibilities e.g. there is a Swedish motorway which charges electric vehicles as they drive.	✗ All have the disadvantage of having an impact on the landscape and causing disturbances to the local environment. Some widespread negative impacts, e.g. in exploiting HEP, large swathes of land are flooded. ✗ **NIMBYism** slows down installation. ✗ Not all renewable energy sources provide the same amount of energy – wind farms take up a lot of space. ✗ Cost is always relative to the price of fossil fuels on world markets and this fluctuates. ✗ Very few countries have climates where renewables might completely replace fossil fuels.

Nuclear power

Nuclear power is the most realistic option for replacing fossil fuels but has disadvantages:

- Nuclear disasters such as Fukushima could happen again.
- Terrorists could target nuclear powered stations.
- Radioactive waste must be disposed of safely, often through **vitrification** and storage underground.

- The technology involved is only accessible for developed countries.
- Construction and decommissioning costs are extremely high.
- Energy security may be compromised if FDI is from foreign TNCs.

Biofuels

In theory, biofuels are a good alternative to fossil fuels. The process of burning biofuels is thought to be **carbon neutral**: CO_2 taken in by a plant during photosynthesis is released during combustion. However, across the globe, forests have been cleared to plant oil palms, oilseed rape, and maize to manufacture biofuels. Brazil is the world's second largest producer of biofuels (biodiesel and bioethanol).

Strengths	Weaknesses
<ul><li>Easy and quick to grow feedstock, especially in tropical countries like Brazil.</li><li>Cash crop produces income for farmers.</li><li>Lower CO_2 emissions than fossil fuels.</li></ul>	<ul><li>Converting land to produce biofuels means food security can be threatened.</li><li>Crops require fertilisers which, if artificial, produce greenhouse gases.</li><li>Can increase rates of deforestation as they are more profitable.</li></ul>
Opportunities	**Threats**
<ul><li>Creates a positive multiplier effect, e.g. investment in roads and irrigation systems.</li><li>Diversifies the rural economy.</li></ul>	<ul><li>Overuse of fertilisers can pollute local waterways and lead to eutrophication.</li></ul>

Radical technologies

Hydrogen fuel cells

- ✓ Water is the only waste product.
- ✓ Great potential for cars and public transport.
- ✓ Prototypes already developed.
- ✗ Separating hydrogen requires large amounts of energy and releases greenhouse gases.
- ✗ Hydrogen is not a primary energy source, only a way of storing energy.

Electric vehicles

- ✓ Useful in urban environments as air pollution is reduced.
- ✓ Slowly becoming cheaper.
- ✓ Reduce noise pollution and greenhouse gas emissions.
- ✗ Recharging infrastructure is lagging.
- ✗ Range is not yet adequate for longer journeys.

Carbon capture and storage

- ✓ Reduces emissions from power stations by up to 90%.
- ✓ Provides a backup source of energy when renewables are not available.
- ✗ Requires significant investments in research, development, and infrastructure, raising costs.

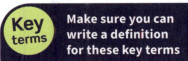

Key terms — Make sure you can write a definition for these key terms

carbon footprint   carbon neutral   net zero   NIMBYism
radical technologies   recyclable energy   vitrification

RETRIEVAL

Learn the answers to the questions below, then cover the answers column with a piece of paper and write as many as you can. Check and repeat.

Questions | ## Answers

	Questions		Answers
1	Name the four groups of key players involved in energy supply and pathways.		TNCs, OPEC, governments, consumers
2	Give the three main reasons why energy security is important for countries.		Economic growth and development, keep energy prices low, helps with social stability
3	If a country can increase the proportion of electricity it produces domestically, what happens to its energy security?		It improves
4	How would improving energy efficiency affect energy consumption per capita?		It will reduce it
5	How can unstable geopolitics affect energy prices?		Energy prices can rise if instability is high
6	What is a consequence of an energy mismatch?		An energy pathway
7	Name the four types of energy pathway.		Pipeline, road and rail, shipping, transmission line
8	There are six factors which make energy pathway disruption more likely. State three of these.		Three from: length / multilateral agreements / piracy / areas which are tectonically active / political instability / extreme weather
9	State one advantage of wind power.		Low running costs once installed or plenty of offshore sites
10	Name three radical energy technologies.		Hydrogen fuel cells, electric vehicles, carbon capture and storage

Put paper here

Previous questions

Now go back and use these questions to check your knowledge of previous topics.

Questions | ## Answers

	Questions		Answers
1	On what time scale does the geological carbon cycle operate?		Long term
2	What pump transfers carbon from the surface to the deep ocean?		Carbonate
3	Which organism sequesters carbon dioxide into the ocean through photosynthesis?		Phytoplankton
4	What is the name of the circulation that transfers dissolved carbon around the oceans?		Thermohaline
5	Which type of rock is carbon found in?		Sedimentary

Put paper here

⚙ KNOWLEDGE

8 Carbon cycle and energy security

8.7 Threats to the carbon and water cycles

Threats from human activity

Increasing demand for food and fuel	Land use change
• Increase in population: 9.2 billion by 2050 (UN) especially in S/SE Asia. • Rising affluence and lifestyles: accounts for more of the demand increase. 1990–2015 world energy use increased by 54% while population increased by 36%. • Changing diet: meat consumption will increase from 37.4 kg per person in 2000 to 52 kg/person in 2050 (FAO). Producing this is more water and energy intensive.	• Deforestation: for agricultural production (for food and biofuels), HEP reservoirs, opencast mining (e.g. of tar sands) and infrastructure. • Conversion of grasslands: for agriculture and biofuel production in LICs and newly industrialising countries (NICs). Happens particularly in tropical grasslands (savannah) as biofuels grow quickly and easily in these habitats. • Afforestation: as a response to **land degradation**. When done correctly, afforestation is a positive land use change.

Impacts of land use change on carbon and water cycles

Impacts on hydrosphere	Impacts on pedosphere	Impacts on atmosphere	Impacts on biosphere
Biofuels need irrigation which impacts long-term stores of water in **aquifers**. Infiltration is reduced as soil is more compacted. Flood peaks are higher and lag time is shorter. Increased sedimentation leaves rivers more prone to flooding.	Soil is more prone to erosion as there is less interception and more surface runoff which reduces the carbon store. Soils lose their structure, become less fertile, and store less carbon. Increased **leaching** of nutrients. Biomass is lost from deforestation so there is less decomposition and organic matter in soils.	Release of CO_2 into atmosphere as soil is ploughed. Reduction in precipitation as transpiration is reduced so there is less water vapour in the atmosphere. Less humid air as less evapotranspiration.	Less photosynthesis so the efficacy of the biosphere as a carbon store reduces. Reduction in leaf litter and biomass, so less carbon added to the soil. Species diversity is lost as agriculture is usually a **monoculture**. Monocultures tend to store less carbon.

Global and regional trends

- Deforestation and grassland conversion for farming is mainly in NICs, LICs and **MINT countries**.
- Afforestation is mainly in HICs. Exceptions are Costa Rica and the 'Green Wall' countries of the Sahel e.g. Ethiopia – this project aims to reforest 3 million hectares by 2030.
- Deforestation occurs at a rate of 18 million hectares a year – equal to 27 soccer fields every minute (WWF).

- In 2015, 30% of all global forest cover had been completely cleared (50% according to the FAO) and 20% degraded.
- In Africa and South America, most forests have halved in area since the 1960s.
- In Indonesia, large areas of forest land have been cut down or burnt to make way for palm oil plantations – demand for this product is increasing significantly.

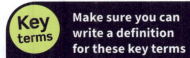

Key terms Make sure you can write a definition for these key terms

acidification   aquifer   coral bleaching CO_2 fertilisation effect
degradation   ecosystem resilience   leaching   MINT country

8 Carbon cycle and energy security

8.7 Threats to the carbon and water cycles

Changes in ocean acidification

- Oceans are carbon sinks that absorb about 30% of the CO_2 produced by humans.
- Oceans are becoming increasingly acidic as more CO_2 is sequestered.
- pH has dropped by 0.1 since the Industrial Revolution (an increase in acidity of about 26%).
- Predictions are for a further drop of 0.06–0.32 by the end of the century.

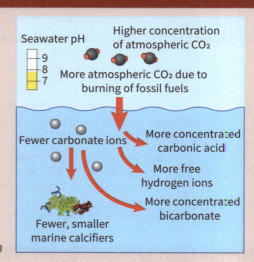

▶ **Figure 1** *Ocean acidification*

Impacts of ocean acidification

- A reduction in **ecosystem resilience** and potential for crossing a threshold is increased beyond a tipping point.
- Ocean acidification impacts could be amplified by other stressors such as thermal expansion, rising sea levels or destructive cyclones.

- The Arctic Ocean is likely to be first affected because of its low pH, threatening cold corals.
- Coral reefs worldwide are at risk of **coral bleaching** from temperature increases and acidification that dissolves carbonate structures. Major ecosystems based on coral reefs could collapse.

Climate change and droughts in the Amazon

There are complex feedback loops in climate–ecosystem relationships.

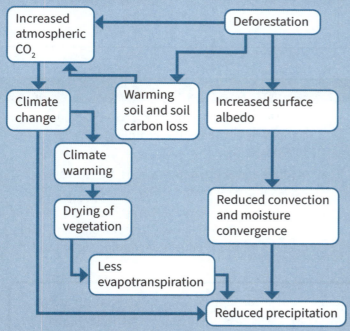

- Tropical forests have protected us against climate change in two ways:
  - They hold roughly a quarter of the world's stored carbon and absorb more CO_2 than they release.
  - A **CO_2 fertilisation effect** that results from increased levels of carbon dioxide in the atmosphere has increased the rate of photosynthesis in plants.
- Deforestation, anthropogenic climate change and shifting climate belts are creating positive feedback loops leading to e.g. more frequent droughts.
- Complexity of interactions makes predicting the future impact uncertain.
- CO_2 emissions from forest fires are now half as great as those from deforestation.
- An ecological tipping point may have been reached as more carbon is now released than stored, meaning ecosystems could collapse and transform to savannah if droughts persist.

8.8 Threats to human wellbeing

Forest loss

The UN describes forests as fundamental to human wellbeing and survival.

- Forests are essential due to the ecosystem services that they provide.
- They are the source of around 80% of global biodiversity.
- Over 1.6 billion people depend on forests, and over 90% of these are the poorest in society.

Forests provide ecosystem services and support human wellbeing:

- Provisioning, e.g. food, water, fuel, generic services
- Regulating, e.g. climate, floods, storms, water purification
- Cultural, e.g. educational, recreational, religious
- Supporting, e.g. soil formation, wildlife habitats, nutrient cycling.

Local impacts	Global impacts
<ul><li>Disruptions to hydrological cycle, e.g. flooding, means loss of property and higher insurance.</li><li>Increased soil erosion which decreases agricultural productivity leading to more food insecurity.</li><li>Loss of traditional herbal-based remedies.</li><li>Displacement of people who are directly reliant on the provisioning services of ecosystems, and the loss of related traditional cultures, e.g. the Efe tribe in central Congo.</li></ul>	<ul><li>Loss of potential cures to major diseases such as cancer that may be found in plant derivatives.</li><li>Loss of regulating services such as carbon dioxide sinks.</li><li>Changes to agricultural production because of shifting **climate belts**.</li></ul>

Forest recovery trends

There is evidence that forest stores are being protected, and even expanded, in some locations. The effects of this can be shown in the environmental Kuznets curve model.

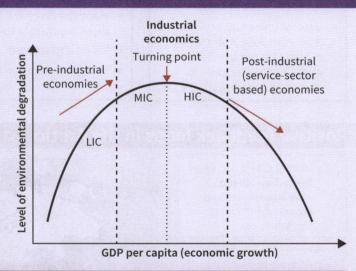

▶ **Figure 1** *Environmental Kuznets curve model*

Positives	Limitations
<ul><li>Suggests a general pattern that, once a tipping point in environmental degradation is reached, policies become more **protectionist**.</li><li>Public opinion and government policy in HICs are focused on sustainable living and restoration of forests.</li></ul>	<ul><li>Many industrialising and middle-income nations are focused on sustainability. For example: in Indonesia, a moratorium banning clearance of forest and peatland for timber or palm oil has reduced rates of forest clearance by 15%.</li><li>Protective legislation in many LICs is combined with bottom-up approaches that are critical for long-term reduction in forest loss.</li><li>Costa Rica tops the **Happy Planet Index** due to government policies around sustainability which benefit the majority.</li></ul>

Key terms Make sure you can write a definition for these key terms

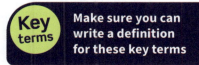

albedo   climate belts   cryosphere   environmental refugees
Happy Planet Index   river regime

8 Carbon cycle and energy security

8.8 Threats to human wellbeing

Rising temperatures cause change and human impacts

Precipiation patterns

- Higher temperatures lead to greater evaporation and surface drying, potentially with more intense and longer droughts.
- However, as the air warms its water-holding capacity increases by around 7% for every 1°C of temperature rise, leading to more precipitation.
- BUT this will not fall evenly across the planet due to shifting weather patterns. It is expected dry areas will become drier and wet areas become wetter.
- Impacts on humans include increased food insecurity, wildfires and flooding.

River regimes, e.g. Yukon

- Higher temperatures mean more evaporation and water vapour in the atmosphere.
- More precipitation falling as rain rather than snow: the amount is predicted to increase by up to 20% by 2100.
- Shifting **climate belts** mean earlier and faster springmelt in high latitudes leading to higher annual peak discharges.
- Melting permafrost means more infiltration in the longer term and lower surface runoff.

Drainage basin water storage

- Increased evaporation rates reduce soil moisture storage more quickly leading to plant stress (unless irrigated).
- Natural lakes and artificial reservoirs lose water which in some places threatens water security/supply for domestic and industrial uses.

Cryospheric stores

- Melting **cryosphere** reduces water storage in glaciers and ice sheets.
- On a global scale, the water cycle is a closed system so decrease in one store leads to increase in another or an increased flux.
- Many rivers are fed by glacier meltwater in spring and, if this decreases, there are threats to human water supply and security.
- Ice-bound wilderness is opened up for tourism.
- Energy exploration is more feasible as ice melts, improving energy security.

Positive feedback loops in the Arctic

1. Melting sea ice decreases the surface **albedo**. More solar radiation is absorbed, increasing sea temperatures.

2. Offshore permafrost melts releasing large amounts of frozen methane.

3. Melting of Greenland ice sheet leads to sea level rise.

2. Arctic Ocean

4. Rising temperatures increase water vapour in the atmpophere which traps more outgoing long wave radiation causing further warming.

5. Temperature of river discharge is higher due to warming land surfaces. This increases sea temperature which amplifies sea ice melting.

▲ **Figure 2** Positive feedback loops in the Arctic

Threats to ocean health

- Acidification is reducing primary producers.
- Fish stocks are declining and distributions changing leading to challenges for fishing communities.
- Tourism and marine biodiversity in countries with coral reefs is threatened.
- Rising sea levels threaten coastal communities and sometimes whole nations, leading to **environmental refugees**.

> **REVISION TIP**
>
> In your revision notes, make clear links between changes to physical systems and human wellbeing.

8.9 Future uncertainties, risks and responses

What does the future hold?

Climate models predict a temperature rise of between 2 °C and 6 °C.

Temperatures are expected to warm more rapidly over the Northern Hemisphere.

The future

Temperature increases are driven by industrialising economies such as India.

Even if all emissions stopped today, temperatures would continue to rise due to the residence time of greenhouse gases in the atmosphere.

Why is future climate change so uncertain?

The role of carbon sinks may change

- Taiga and boreal forest may migrate north and be able to store more CO_2.
- Tropical rainforests are already at carbon capacity and storage may reduce if droughts intensify.
- Although phytoplankton, sea grasses and algae are predicted to increase, oceans may be less able to store carbon because:
  — warming reduces the solubility of CO_2
  — the efficiency of the biological pump taking dissolved carbon to the ocean floor may fall.

Human factors may change

- Energy mixes may change and increasingly decouple from fossil fuels.
- Land use changes may slow down or be reversed due to new technologies or government policy.
- Future economic growth is uncertain. Economic downturns may slow manufacturing.
- New resource finds (e.g. of fracking sites in the USA) may impact emissions.
- New technologies in energy production may have an impact.

Feedback mechanisms and tipping points could amplify changes

Positive feedbacks

These include:
- Thawing of permafrost in the Arctic could release more carbon dioxide and methane, increasing the rate of climate change.
- Waterlogged peat deposits may dry out as water tables fall and release more methane to the atmosphere.

Negative feedbacks

These include:
- Increase in plant growth: with more CO_2 in the atmosphere, the biosphere increases its capacity as a carbon sink.
- Increase in cloud cover caused by more water vapour held in warmer air could reflect more solar radiation, reducing heating.

Tipping points

These include:
- Atlantic thermohaline circulation could collapse due to changes in water density and salinity caused by melting ice sheets. This might affect Northern Hemisphere temperatures.
- Dieback of tropical forests due to shifting climate belts leading to increased drought and increasing release of CO_2.

8 Carbon cycle and energy security

8.9 Future uncertainties, risks and responses

Adaptation strategies

Strategy	Description	Risks and costs
Water conservation and management	It is essential to reduce water consumption per capita through: • **Drip irrigation** and use of GIS and remote sensing to target irrigation of crops. • Recycling sewage water for agricultural use. • Reusing **grey water** in washing machines and toilets. • Metering consumers and charging 'real-value' prices to reduce domestic consumption.	• Potentially greater threats to public health. • Retrofitting homes or introducing new technologies costs a great deal of money.
Resilient agricultural systems	Food production needs to adapt to more extreme climate by: • Using no-ploughing approaches to reduce soil erosion during flood events and release less carbon to the atmosphere. • Reducing use less artificial fertiliser which is a source of greenhouse gases. • Breeding of drought-resistant crops or crops that can withstand higher wind speeds to maintain yields. • Re-establishment of hedgerows to improve biodiversity and increase photosynthesis.	• The farming community may be resistant to new ideas if profits are at risk in the short term. • Food prices may rise.
Land-use planning and flood risk management	• **Rewild** rivers by allowing them to return to natural courses. • Limit development on flood plains, leaving land as playing fields, parks and fields to increase infiltration rates.	• There are other pressures such as housing shortages. • Changes to land use in one place always have unintended consequences somewhere else.

Mitigation strategies

Solar radiation management

- Use of climate engineering such as spraying aerosols into the upper atmosphere or brightening clouds to reflect the Sun's rays.

Carbon taxation

- The carbon tax price floor set a minimum price companies had to pay to emit carbon dioxide. It was unpopular with both industry and environmental groups and had a debatable effect on emissions.
- Lower road taxes for low-carbon-emitting cars in the UK were scrapped in 2015.
- Introduction of low emissions zones in many UK cities, such as ULEZ in London, with penalties for polluting vehicles.

Renewable switching

- Sweden is leading the world with switching to renewable energy sources to lower carbon per capita even though its consumption per capita is high.
- Renewables provide intermittent electricity, while fossil fuels provide the continuous power essential for infrastructure.

Energy efficiency

- The UK's Green Deal scheme encouraged energy-saving improvements to homes, such as efficient boilers and lighting and improved insulation.
- Germany's economy has grown even though energy consumption has dropped due to energy efficiency measures.

Afforestation

- Tree planting in the UK is increasing, helping carbon sequestration. It is often coupled with offsetting schemes.
- The Big Tree Plant campaign encourages communities to plant 1 million new trees, mostly in urban areas.

Carbon capture and storage (CCS)

- Canada's Boundary Dam is the only large-scale working scheme which captures CO_2 from a coal-fired power plant.

Agreements and key players

Globally, agreement is difficult as countries put national interests first.

Paris Agreement 2015

The Paris Agreement is the most significant agreement. It aims to:

- limit global temperature rise to 1.5°C
- strengthen adaptation and mitigation strategies
- provide support to initiatives in the developing world.

National governments

National governments have a range of concerns. They:

- are uncertain of the best way of achieving progress
- don't want to hamper economic growth
- don't want to lose votes due to levies or taxes to cut carbon emissions.

Individual citizens

Individual citizens have different viewpoints. For example:

- people in countries at risk of sea level rise are concerned
- if jobs are threatened by government climate policies, people might be resistant.

Key terms
Make sure you can write a definition for these key terms

adaptation strategy   carbon capture and storage (CCS)
drip irrigation   grey water   mitigation strategy
Paris Agreement   rewilding

Learn the answers to the questions below, then cover the answers column with a piece of paper and write as many as you can. Check and repeat.

Questions | Answers

	Questions	Answers
1	What three factors are driving an increasing demand for food and fuel?	Population increases, rising affluence, changing diet
2	State the three types of land use change which drive changes in the carbon cycle.	Deforestation, afforestation, grassland conversion
3	As the ocean absorbs more carbon dioxide, is it becoming more acidic or alkaline?	More acidic
4	Which type of marine ecosystem is being bleached as a result of oceans absorbing more carbon?	Coral reefs
5	Name four groups of ecosystem services.	Regulating, supporting, provisioning, cultural
6	What is the name of the environmental model used to show the relationships between degradation and economic development?	Kuznets curve
7	Rising temperature will most likely cause dry areas to become drier and wet areas to become wetter. True or false?	True
8	As a result of rising temperature, what will happen to cryospheric stores of water?	They will decrease
9	State two mitigation strategies.	Two from: carbon capture and storage / renewable switching / afforestation / energy efficiency / carbon taxation
10	Name the global agreement which aims to limit future temperature rise to 1.5°C.	Paris Agreement

Put paper here

Previous questions

Now go back and use these questions to check your knowledge of previous topics.

Questions | Answers

	Questions	Answers
1	How would improving energy efficiency affect energy consumption per capita?	It will reduce it
2	Name four types of energy pathway?	Pipeline, road and rail, shipping, transmission line
3	How can unstable geopolitics affect energy prices?	Energy prices can rise if instability is high
4	Differences in temperature and salinity cause a difference of what in the ocean?	Density
5	Name four greenhouse gases.	Carbon dioxide, methane, nitrous oxide, fluorinated gases

Put paper here

PRACTICE

Exam-style questions

1   Explain the geological processes which influence the amount of carbon in the atmosphere.   **(6)**

2   Explain the formation of fossil fuels such as oil.   **(6)**

3   Explain the significance of photosynthesis in maintaining planetary health.   **(6)**

4   Explain how carbon is transferred from the atmosphere to the oceans.   **(6)**

5   Explain how energy consumption per capita may change as a country develops economically.   **(6)**

6   Explain the reasons for the variations in emission of CO_2 from global land use change.   **(6)**

7   Study **Figure 1**. Explain how increased atmospheric CO_2 can cause positive feedback loops in tropical rainforest ecosystems.   **(6)**

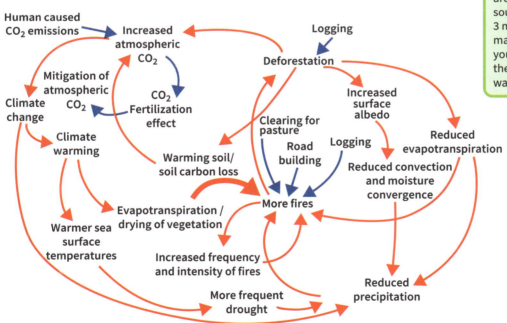

▲ *Figure 1 Positive feedback loops in tropical rainforest ecosystems*

8   Study **Figure 2**. Explain why projected future global surface warming may be uncertain. **(6)**

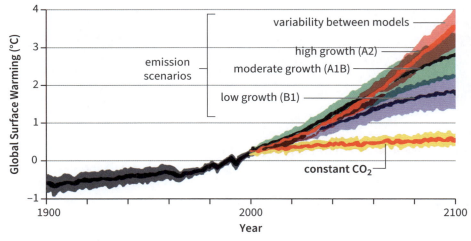

▲ *Figure 2 Graph of global surface warming*

9   Explain why the energy mixes of countries vary. **(8)**

10   Explain how political tension and conflict may disrupt energy pathways. **(8)**

11   Explain how a national commitment to net zero can lead to changes in the energy mix of a country. **(8)**

12   Explain how feedback loops and tipping points can contribute to future climate change. **(8)**

13 Assess the impacts of climate warming on the carbon cycle of the taiga biome shown in **Figure 3**. **(12)**

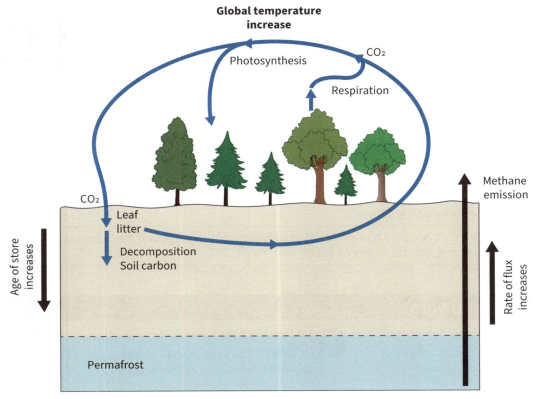

Global temperature increase

Photosynthesis

CO_2

Respiration

Methane emission

CO_2

Leaf litter

Decomposition Soil carbon

Age of store increases

Rate of flux increases

Permafrost

▲ **Figure 3** *Taiga biome found in the Arctic*

14 Assess the potential impacts of human activity on the water and biological carbon cycles. **(12)**

15 Assess the implications for human wellbeing of the degradation the carbon and water cycles. **(12)**

16 Evaluate the view that the exploitation of unconventional fossil fuels is inevitable. **(20)**

17 Evaluate the view that renewable energy will not be able to replace fossil fuels as primary energy sources. **(20)**

18 Evaluate the view that mitigation strategies can significantly reduce future global warming. **(20)**

> **EXAM TIP** 🎯
>
> Examiners want to see judgements about the 'relative significance or importance of factors.' One way to achieve this is to start your paragraphs with a sentence such as 'The most significant potential impact is ...'

⚙ KNOWLEDGE

9 Superpowers

9.1 Defining superpowers

Levels of geopolitical power

Superpower	A country that has significant global strength. It exerts its political, economic, military, and cultural influence beyond its borders, making it a dominant global force, e.g. USA.
Emerging power	A country that has the potential to be a superpower in the future. It has significant economic power and is increasing its political, military, and cultural influence around the world, e.g. China.
Regional power	A country that has significant influence within a specific geographic region, perhaps within a continent, e.g. Brazil.
Hyperpower	A country that not only has all the characteristics of a superpower but remains unchallenged, e.g. USA 1991 to 2010.

Characteristics of superpowers

1 Economic power: high GDP per capita; strong, stable, and diversified economy; sustainable economic growth; ownership of TNCs which are significant global players/brands; stable currencies; trade partnerships have a global reach.

2 Military strength: significant military capability (land, sea, air), advanced weapons, and other technology (e.g. nuclear, intelligence, cyber operations) that can be used to defend its territory but also to threaten or invade other countries; military expertise may be offered to other countries (military and civilian scenarios); maintains a presence in multiple world regions.

3 Demography: large population with a good standard of living (health/education/housing/resource access); sustainable population growth.

4 Political influence: holds influence on the global stage; central role in IGOs such as UN, WTO, and can influence major global events and the decision-making of other countries, often to the advantage of the superpower; extensive diplomatic network; acts as mediator in regional disputes.

5 Cultural influence: country has values, behaviours, or beliefs (ideology) which it imposes on others or that other countries want to follow; language, media, arts, fashion, and entertainment are exported to other countries; globally recognised universities attract students from around the world.

6 Resource access: access to and control over a wide range of essential resources, including energy and minerals, either within the country or through influence on other nations; the country's own resource base, its use, and its development for trade is influenced by physical factors such as land surface area, relief, accessible coastline, climate, soils, and vegetation.

Some characteristics have increased in importance as the twenty-first century progresses:

- Technologically advanced: high levels of research and development (R&D) in fields such as scientific innovation, pharmaceuticals, artificial intelligence, and cyber and space technology.

- Resilience: can withstand and recover from significant challenges, e.g. disasters, pandemics, climate change, conflicts, and economic crises, without weakening their global influence.

> **REVISION TIP**
>
> Try to think of a mnemonic to help you remember the different characteristics that are used to identify superpowers.

Mechanisms for maintaining power

- Countries have a mix of strategies to help them maintain and gain power, particularly in regions of the world that have strategic geopolitical importance or key resources.
- Globalisation has helped superpowers, emerging powers, and regional powers exert their influence. 'Detached' countries are less likely to take advantage of this mechanism.
- It can be argued that soft power has gained more importance than hard power over time, although the impacts of **soft power** can take longer to have an effect than those resulting from **hard power**.
- The influence of soft power has developed in the twenty-first century through the internet (including smartphone technology) and the growth of TNCs and their ability to earn more than some countries' economies.

REVISION TIP

Draw two columns and list the differences between hard and soft power.

Hard power ←——— Economic power ——→ Soft power

Military force
- Action or threats by land, sea, air forces
- Military bases in other countries
- Nuclear weapons
- Military alliances (NATO)

Economic force
- Sanctions
- Embargoes
- Boycotts

Trade and aid
- TNC influence including FDI
- Trade blocs and alliances, e.g. EU/NAFTA
- Bilateral/multilateral trade agreements

Culture
- Export of cultures through language, media including Internet, streaming, TV, music, films, social media, sport
- Spread of TNC branding

Ideology
- Assimilation of ideas and beliefs by other countries

Mechanisms for maintaining power have changed over time

USA, USSR, Nazi Germany, Japan

USA increases its economic growth and global influence; Nazi Germany, USSR, Japan exert military power.

USA

Dominant superpower – hyperpower.

USA, China plus EU and emerging/regional powers

USA and China have the most significant economic influence and military power. Emerging continental powers, e.g. Russia, India, Saudi Arabia, and Brazil.

| 1850–1920s | 1920s–1945 | 1945–1991 | 1991–2010 | 2010–2023 | Beyond 2023? |

British Empire

UK controls up to 25% of the land area and governs up to 25% of the global population.

USA and USSR

Cold War bi-polar world.

USA, China, EU

USA maintains its superpower status. China's influence rises beyond just its economy. The EU expands to 28 nations with a population of over 500 million (pre-Brexit 2020).

KNOWLEDGE

9 Superpowers

9.1 Defining superpowers

China's Belt and Road Initiative (BRI)

- China's BRI is a project that develops modern infrastructure to promote international trade.

- It consists of the 'belt' (overland transport routes such as roads, rail, airports and energy infrastructure) and the 'road' (maritime shipping routes and ports).

- Over 140 countries, with 40% of global GDP and 60% of global population, will be directly influenced by the BRI.

- Through the BRI, China exerts 'soft' rather than 'hard power' by offering support and other assistance to partnering countries.

- The BRI allows goods to be traded and aid, political support and other assistance to be provided to partnering countries.

- The BRI has been used to promote medical research and delivery of healthcare supplies and expertise to partner countries (Health Silk Road), and to improve digital connectivity on a regional and local scale (Digital Silk Road).

- However, some argue that the reliance of partner countries on finance from China to develop projects leaves them open to dependency, obligation and debt to China.

- BRI projects have given China economic and political advantages. Partner countries also benefit but some may experience problems.

- Future developments include further cross-border projects linked to health and digital technology.

▲ **Figure 1** The Belt and the Road

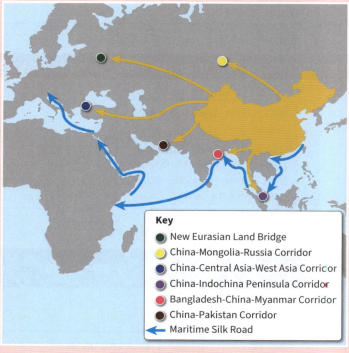

▲ **Figure 2** BRI project locations

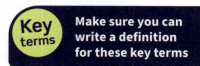 **Make sure you can write a definition for these key terms**

Belt and Road Initiative (BRI)   emerging power   hard power
hyperpower   regional power   soft power   superpower

9.2 Changing patterns of power

The concept of polarity

Uni-polar: one major global power exists and tends to dominate the global system.

Unipolar

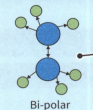

Bi-polar: two dominant powers which can be on the brink of conflict.

Bi-polar

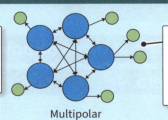

Multi-polar: three or more powers competing with each other for global dominance.

Multipolar

Key
⟶ Connections with/influence over

Uni-polar world: the British Empire

- The **British Empire's** growth was based on colonisation. The Empire's economic interests were protected by the British Army and the Royal Navy.
- From 1850, colonisation expanded inland. Trade expanded. The Industrial Revolution increased UK manufacturing of products exported to colonies.
- Colonial administrators offered local leaders support and protection to gain and help maintain direct control.
- Acculturation occurred as British values and customs spread to Britain's colonies.

Multi-polar world: 1919–1939

Between the First and Second World Wars, the rise in power of the USA, the USSR, Japan, and Nazi Germany challenged Great Britain's global position, creating regional spheres of influence.

Bi-polar world: Cold War era

- Between 1945 and 1970, the majority of Britain's colonies took back independence, e.g. India in 1947.
- After 1945, a bi-polar system emerged with the USA (capitalist) and USSR (communist) as superpowers supported by their respective allies.

- The **Cold War** was a period of political tension, including a nuclear arms race, proxy wars, intelligence gathering, and the space race.
- Multi-faceted indirect control became increasingly important.

	USA	USSR
Political	Democracy – federal republic of 50 states with a president elected every four years	Dictatorship – single party state controlling Russia and other Soviet Socialist Republics in Eurasia
Economic	Free trade with few barriers; privately owned TNCs; supply, demand, and prices linked to the market	State-owned businesses meeting demands of the command economy; central government sets levels of production and prices
Military	Nuclear weapons; NATO alliance with strong presence in Western Europe; powerful navy and air force	Nuclear weapons; Warsaw Pact alliance with strong presence in Eastern Europe; large army; military spending in 1980 was more than the USA
Cultural	Influenced cultures around the world by spread of pop culture; consumer society	State-controlled propaganda and media; exchanges showcased USSR culture; focus on e.g. ballet, classical music, sculpture, and sport

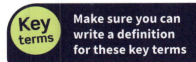

Key terms Make sure you can write a definition for these key terms

bi-polar   British Empire   Cold War   geopolitical stability
hegemony   multi-polar   neo-colonialism   uni-polar

9 Superpowers

9.2 Changing patterns of power

Neo-colonial era

Neo-colonialism is the use of soft power to obtain and sustain influence in developing nations.

Mechanisms used to promote neo-colonialism include:

- TNCs and FDI: used to develop economic relations between countries but the investor is the dominant force, leading to dependency.
- Tied aid: development assistance given to a country but with certain conditions that favour the donor country.
- Debt burden: unfavourable terms and conditions on debts or loans (including debt servicing), leading to financial dependency.

- Trade: LICs exporting lower-value commodities to HICs but importing higher-value goods from them, leading to economic dependency; global trade agreements may favour superpowers.
- Political interference: support for governments in LICs that favour the dominant country's interests; election interference; superpowers use dominant position in IGOs to advance their interests.
- Cultural influence: continued spread of 'western' values and consumer culture through media , including the internet.

The rise of China

- The end of the Cold War left the USA leading a uni-polar world, and its economic, cultural, political, and military strength remained unrivalled.
- China is now emerging as a rival to the USA's **hegemony** due to its rapid economic growth and military expansion.

- China exerts its economic influence through its Belt and Road Initiative (BRI) and investment in resource extraction and infrastructure, particularly in Africa.
- As education standards increase, China faces problems due to its lack of democracy and censorship.

Geopolitical stability and risk

Each type of polarity has different levels of **geopolitical stability** and risk.

Geopolitical stability	System	Risk
One dominant power is able to act to limit the potential for a global war. It can exert significant military, economic, political, and cultural influence to maintain its power.	Uni-polar	Other nations may challenge the dominant power, causing instability.
Although conflict exists in other countries, the equality between the two major poles limits the chances of a major conflict between them.	Bi-polar	If peace between the two poles is threatened, and diplomacy unsuccessful, then the bi-polar system will become unstable and the risk of war is increased.
The changing alliances, partnerships, and power balance between the poles limits the opportunities for conflict.	Multi-polar	Changing relationships may become more unpredictable, increasing the chance of conflicts, particularly regionally.

> **REVISION TIP**
>
> Make a one-minute audio recording to explain how geopolitical stability changes as a country's power increases or decreases.

9.3 Emerging superpowers

The economic importance of emerging countries

- Some emerging countries have increasing global influence, due to changing economic structure and economic growth, leading to increased levels of development.

- There has been slower economic growth in the USA.
- Ageing populations may affect the economic influence of the EU and Japan.

BRICS

- An intergovernmental organisation (IGO) comprising Brazil, Russia, India, China (2006), South Africa (2010), Egypt, Ethiopia, Iran, Saudi Arabia, United Arab Emirates (2024).
- Could challenge US dominance of the global economy by 2050 and are seen as a counterbalance to the Western-led world.

G20

- An IGO, which promotes international economic cooperation.
- Members have political stability and influence as well as a substantial economy.
- Other emerging economies in the G20 (for example, Saudi Arabia) have increasing regional and global influence.

Global environmental governance

- China and India are among the emerging nations to sign up to the UN Framework Convention on Climate Change.
- Emerging economies such as the BRICS have a role in tackling issues such as water pollution, deforestation, habitat loss, and the impacts of oil, gas, and mineral extraction.
- They also are working towards the UN Sustainable Development Goals related to the environment.

Strengths and weaknesses of emerging powers

Brazil

Strengths

- Large amounts of natural resources
- Leading exporter of agricultural products, e.g. soybeans
- Strong regional influence in South America
- Strong renewable energy sector, including HEP and biofuels
- Hosted global sporting events (2014 Men's Football World Cup, 2016 Olympics and Paralympics)

Weaknesses

- Internal inequalities in income (Gini Coefficient 52.9) and social disparities in access to health and education
- Volatile economy
- Political instability and corruption
- Environmental issues, e.g. deforestation
- Second largest armed forces in the Americas (after USA) but regional rather than global focus
- High levels of crime, particularly in urban areas
- Protectionist trade policies

Russia

Strengths

- Geopolitical influence
- Nuclear weapons and large number of armed forces
- Energy rich (oil, natural gas)
- Permanent UN Security Council seat

Weaknesses

- Economy dependent on energy
- Sanctions and isolation due to war on Ukraine
- Shrinking and ageing population
- Human rights issues

9 Superpowers

India

Strengths

- Large, youthful population
- Large potential market
- Robust economic growth
- Large number of active military personnel
- Nuclear power and innovator in space technology
- Educated workforce
- World's largest democracy

Weaknesses

- Poverty and inequality – although poverty is declining, 10% of population are in extreme poverty
- Investment needed in transport and energy infrastructure
- Geopolitical tensions with Pakistan and China
- Large agricultural sector exposed to climate change

China

Strengths

- Robust economic growth
- Strong manufacturing and exports; global trade partnerships
- Large number of active military personnel and investment in military resources
- Rising incomes and aspirations of middle classes
- Established economic links with numerous countries through FDI and other projects
- Educated workforce
- Holds world's largest foreign reserves
- Technological advances – 5G, AI, quantum computing, renewable energy
- Stable government

Weaknesses

- Human rights issues
- Environmental challenges, e.g. air, water pollution
- Ageing population – one child policy removed in 2016 but its legacy remains
- Trade disputes, e.g. with USA
- Geopolitical tensions, e.g. over South China Sea

South Africa

Strengths

- Most developed economy in Africa with strong regional trade links
- Rich in mineral resources, including gold and platinum
- Political transition from apartheid to a multiracial society under Nelson Mandela earned international support
- Biodiversity

Weaknesses

- Highly unequal society, including income inequality (Gini coefficient 63) due to the legacy of apartheid
- High levels of unemployment
- Corruption
- Skills gap due to disparities in access to education and training
- High levels of crime and violence in some areas
- High adult prevalence of HIV/AIDS
- Political uncertainty

Key terms Make sure you can write a definition for these key terms

BRICS   Dependency Theory
G20   Modernisation Theory
World Systems Theory

REVISION TIP

For each BRICS country, write a line explaining why it is important to the global economy or global politics.

Modernisation Theory: Rostow (1960), USA

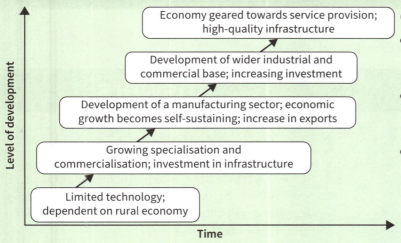

Level of development

Economy geared towards service provision; high-quality infrastructure

Development of wider industrial and commercial base; increasing investment

Development of a manufacturing sector; economic growth becomes self-sustaining; increase in exports

Growing specialisation and commercialisation; investment in infrastructure

Limited technology; dependent on rural economy

Time

▲ **Figure 1 Modernisation Theory** *asserts that countries go through five stages as they develop.*

Critics argue that:

- It focuses on economic development rather than political, cultural, or environmental progress.

- Countries may not follow a similar path to develop, spending different timeframes on each stage or even missing out stages.

- It is based on 'western' development and may not account for other factors that may be more important in non-'western' societies.

REVISION TIP

Remember that these three development theories can be used to help explain changing patterns of power.

Dependency Theory: Frank (1966), Germany

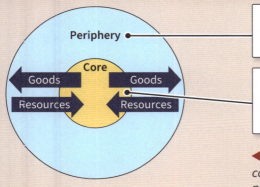

Periphery

Core

Goods   Goods

Resources   Resources

The periphery exports raw materials and cheaper-value products to the core, supplies a lower-cost workforce, and is exposed to debt servicing obligations, weakening the opportunity to invest in its own economy and society.

The core controls the periphery through FDI, economic and military aid, dictating prices for primary products, and dominating decision-making in IGOs such as the IMF and WTO. Colonisation was an a significant factor in the creation of this system and the system is driven by the capitalist world economy.

◀ **Figure 2 Dependency Theory** *argues that the dependency of developing countries (periphery) on more economically developed countries (core) prevents the periphery from making economic progress.*

Critics argue that:

- Emerging economies, including NICs (newly industrialising countries) such do not fit into the model as they are neither in the core nor in the periphery.

- The theory presents a polarised view, making it difficult for countries to transition from the periphery to the core.

World Systems Theory: Wallerstein (1974), USA

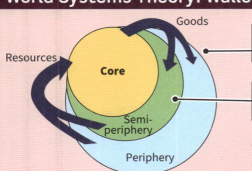

Goods

Resources

Core

Semi-periphery

Periphery

Countries can transition between the core, semi-periphery and periphery over time.

The semi-periphery provides resources, goods and services to the core and also has a dominant influence over the periphery.

◀ **Figure 3** *Wallerstein adds hierarchy to Dependency Theory by adding semi-peripheral regions (including emerging economies) to the model.*

Critics argue that:

- It oversimplifies relationships between different countries.

- By focusing on economic factors, it ignores social, cultural, political, and environmental factors.

Learn the answers to the questions below, then cover the answers with a piece of paper and write as many as you can. Check and repeat.

Questions | Answers

1 What is a hyperpower?

Put paper here

A country that has all the characteristics of a superpower but remains unchallenged

2 What is an emerging power?

Put paper here

A country that has the potential to be a superpower in the future; it has significant economic power and is increasing its political, military, and cultural influence around the world

3 What is a regional power?

Put paper here

A country that has significant influence within a specific geographic region, e.g. within a continent

4 What characteristics can be used to define superpowers?

Economic, political, military, cultural, demographic, access to natural resources

5 What is hard power?

The use of military or economic force to gain influence

6 What is China's Belt and Road Initative?

Put paper here

A project that develops modern infrastructure to promote international trade

7 What is a uni-polar system?

A geopolitical system where one major global power exists and tends to dominate the global system

8 What is a bi-polar system?

Put paper here

A geopolitical system with two dominant powers that can be on the brink of conflict

9 What is acculturation?

The dominance of a new culture over the indigenous one

10 What is neo-colonialism?

Put paper here

Use of soft power (economic, political, and cultural) to gain and maintain influence in a country

11 State two factors that have led to China's emergence as a rival to the USA.

Put paper here

Two from: rapid economic growth / military expansion / BRI / investment in resource extraction and infrastructure

12 Name the BRICS countries.

Put paper here

Brazil, Russia, India, China, South Africa, Egypt, Ethiopia, Iran, Saudi Arabia, United Arab Emirates

13 Name Rostow's five-stage model of economic development.

Put paper here

Modernisation Theory

14 Name Franks's model showing relationships between the developed (core) and developing (periphery) world.

Put paper here

Dependency Theory

15 Name Wallerstein's model.

World Systems Theory

 # KNOWLEDGE

9 Superpowers

9.4 The global economy

The influence of IGOs

Superpowers attempt to control the global economy through different **intergovernmental organisations (IGOs)**.

> **REVISION TIP**
>
> For each IGO, state one way in which it is influenced by superpowers.

World Bank

Founded 1944.

- Supports **capitalism** through loans to developing countries for development projects.
- Funded by financial contributions from member countries and by issuing bonds to investors.
- Provides long-term assistance.
- Aims to end extreme poverty and promote shared prosperity.

International Monetary Fund (IMF)

Founded 1944.

- Provides loans to countries with balance of payments and debt problems.
- Funded by quota contributions from members.
- Short- to mid-term financial assistance.
- Aims to maintain economic stability.

World Trade Organization (WTO)

Founded 1995.

- Operates a global system of trade rules and regulations, promotes **free trade**, helps members negotiate agreements, and resolves trade disputes between nations.
- Aims to use trade to raise living standards and create jobs.

World Economic Forum (WEF)

Founded 1971.

- Not-for-profit foundation linking the public and private sectors.
- Group of political, business, academic, and other leaders who influence global issues.
- Mostly funded by partner companies.
- Hosts annual Davos meeting where political, academic, and business leaders discuss key global issues.

TNCs

- TNCs have become economically dominant in the global economy.
- While many top TNCs are still 'Western' (especially American), TNCs from other countries (especially China) are increasingly influential. Asian TNCs are growing.
- Most TNCS are publicly traded – shares in companies are owned privately by different shareholders.
- The brand is often well-known outside the country of origin.
- In some emerging superpowers (e.g. China), many TNCs are either state owned, or the government is a majority shareholder. This gives governments more control over how the company is run and where the profits go to. The brand may not be well known outside their country of origin.

TNCs in the USA

- Apple
- Microsoft
- Nike
- Amazon
- Walmart

TNCs in China

- State Grid Corporation of China
- Huawei
- Sinopec
- Sinochem
- Alibaba Group
- NIO

 Key terms

Make sure you can write a definition for these key terms

capitalism   free trade   intergovernmental organisation (IGO)
International Monetary Fund (IMF)   patent   trade pattern
westernisation   World Bank   World Economic Forum (WEF)
World Trade Organization (WTO)

9 Superpowers

9.4 The global economy

The five largest TNCs in terms of revenue (2023)

Name (Country)	Publicly traded or state owned	Business type	Revenue ($billion)	Employees
Walmart (USA)	Public	Retail	611	2,100,000
Saudi Aramco (Saudi Arabia)	Public (Saudi government majority shareholder)	Oil, gas, minerals	603	70,496
State Grid (China)	State	Energy	530	870,287
Amazon (USA)	Public	Internet services/ retailing	513	1,541,000
China National Petroleum (China)	State	Oil and gas	483	1,087,049

Factors in the dominance of TNCs

TNCs invest in researching and developing new products and ways to make their businesses more efficient.

They hold **patents** for new inventions which gives them the sole right to make, use, or sell their new products.

Technology

Patents, along with trademarks and copyright, are protected by intellectual property (IP) law, made by governments and promoted by the UN's World Intellectual Property Organization (WIPO).

TNCs can control vital resources, e.g. energy, raw materials, and IP. This control influences trade as a TNC can make it more difficult for competitors to access resources.

TNCs produce large volumes of high-quality products using more streamlined production methods, and have efficient supply chains which allow for easier imports and exports between the countries where the TNC operates.

Trade patterns

TNCs have more capacity to recover from external shocks, e.g. pandemics, energy crises.

> **REVISION TIP**
>
> Draw two columns, one for 'Technology' and one for 'Trade patterns'. Write down as many factors as you can remember in each column.

Global cultural influence

- The dominance of North American and western-European TNCs has contributed to the '**westernisation**' of cultures around the world.
- Developments in technology have contributed to the globalisation of ideas and cultures.
- Films, television, music, and other forms of 'western' culture, are available around much of the world.
- Globalisation also applies to food – international food is available widely in the UK and other 'western' countries.

9.5 International decision-making

Crisis response

- When there is an international crisis, countries ask other countries and IGOs for assistance. Who they ask for help depends on the nature of the crisis, their relationships and alliances with other countries or groups of countries, and their membership of IGOs.

- The ability of countries to assist depends not only on their political relationships with other countries but also on their own access to money, technology, resources, and expertise.

- The UN and other IGOs often take the lead in a global crisis. Superpowers and emerging nations will assist IGO responses. Individual governments may also provide aid and assistance.

Response to conflict

- Powerful countries can have a significant influence in global politics. The USA has sometimes been referred to as the 'global police' because of its involvement in international conflicts and peacekeeping, particularly since the Second World War.

- The USA has acted unilaterally, with other allies (e.g. NATO), and as part of the **UN Security Council** to influence foreign affairs. This has included maintaining and strengthening political and economic ties with other countries, and protecting their economic and geopolitical interests in other parts of the world.

- Evolving US foreign policy, changing attitudes to the USA's involvement in recent conflicts (e.g. the 2003 invasion of Iraq), and the rise in power of emerging economies will influence the USA's future 'global police' role.

▲ *Figure 1* *The UN Security Council chamber is located in the UN Headquarters in New York, USA*

Climate change

The UN Framework Convention on Climate Change (UNFCCC) is the main international treaty to support the global response to the threat of climate change.

Response to climate change

As superpowers are significant contributors to climate change, there is pressure for them to act on this issue.

REVISION TIP

Remember, you need to know how superpowers act in times of crisis, conflict, and climate change.

9 Superpowers

9.5 International decision-making

Alliances

- Countries create alliances based on similarities of interests. By being part of an alliance, countries gain support of other nations and strengthen their global influence and geo-strategic position.

- Alliances increase interdependence between countries as they have to work together towards their common goals.

Military alliances

North Atlantic Treaty Organisation (NATO)

- **NATO** members cooperate on defence and security with the long-term aim of preventing conflict.

- If diplomacy fails, each member state agrees to defend others that are attacked by a country outside the group. Member states also guarantee not to initiate a conflict with another NATO member.

- In the Cold War, it acted as a balance to the threat of the Soviet Union and its allies. Recent expansion of NATO has led to tensions with Russia.

Australia, New Zealand and United States Security Treaty (ANZUS)

- **ANZUS** member states cooperate on peace and security issues in the Pacific Region.

- New Zealand has been partially suspended from ANZUS due to its nuclear-free policy, as Australia and the USA have a pro-nuclear policy.

- ANZUS has been put under further strain by the AUKUS pact (2021) between Australia, the USA, and the UK through which the USA and the UK will help Australia to obtain nuclear-powered submarines.

Economic alliances

European Union (EU)

- The **EU** is one of the largest trade blocs in the world.

- Promotes free movement of people, goods, services, and capital between member states.

- The UK was a member of the EU until 2020.

United States–Mexico–Canada Agreement (USMCA)

- NAFTA was a free trade zone between the USA, Canada, and Mexico.

- In 2008, the USA abolished all tariffs and quotas on exports to Mexico and Canada. In 2020, NAFTA was replaced by the United States–Mexico–Canada Agreement **(USMCA)**.

- The agreement is an important part of USA–Mexico relations. In 2022, Mexico was the USA's second largest trading partner in terms of both imports and exports.

Association of Southeast Asian Nations (ASEAN)

- **ASEAN** aims to promote economic growth and social development within the region.

- Member states work together to maintain political, economic and environmental stability.

- Member states then benefit from economic development, lower poverty levels, and reduced socio-economic disparities.

- Through ASEAN, member states have more global influence.

Environmental alliances

Intergovernmental Panel on Climate Change (IPCC)

- The **IPCC** is the UN organisation responsible for assessing the science linked to climate change.

- Experts from its 195 member states review scientific research and produce Assessment Reports about the impacts and risks of, and potential responses to, climate change.

The United Nations (UN)

- Established in 1945, the UN is the world's most influential IGO.
- Its 193 member states cooperate on a range of global issues including peace and security, human rights, aid, and sustainable development.
- To carry out its work, the UN system includes a number of bodies, funds, programmes, and specialised agencies.

REVISION TIP

Practise writing an explanation of why the UN is important for global stability.

Key parts of the UN system

United Nations Security Council

- Responsible for maintaining international peace and security.
- Five permanent members (USA, UK, Russia, France, and China), who have the power to veto resolutions. Ten temporary members have seats allocated on rotation based on geographic region.
- All UN member states have to follow decisions made by the Council.
- Critics argue that this system gives more influence to the world's most powerful nations.

International Court of Justice (ICJ)

- The main legal body of the UN.
- Uses international law to settle legal disputes on behalf of countries, and offers legal advice to other UN bodies.
- Located in the Hague (Netherlands), it is the only one of the six main UN bodies not to be in New York (USA).

UN peacekeeping missions

- Aim to help countries achieve long-term peace and stability.
- UN peacekeepers are a neutral force whose role is to protect civilians in conflict regions and help to maintain ceasefires.
- They are part of the political process that aims to maintain security in politically unstable regions.
- Member states provide troops and police who work with civilian peacekeepers to uphold decisions made by the UN Security Council.
- Recent peacekeeping missions include the Middle East, Cyprus and the Democratic Republic of Congo.

Climate change conferences

- Conferences of the Parties (COPs) are the decision-making body of the UNFCCC.
- Member states usually meet every year to review progress towards limiting climate change.
- Member states will debate and negotiate climate change issues, leading to agreements and treaties, e.g. the Paris Agreement came from COP 21 (2015).
- COPs can try to refine targets set, e.g. at COP26, member states debated the reduction of the 2 °C limit on average global temperature increase agreed in Paris to 1.5°C.

Key terms

Make sure you can write a definition for these key terms

ANZUS   ASEAN   EU   ICJ   IPCC   NATO
UN peacekeeping mission   UN Security Council   USMCA

9 Superpowers

9.6 Superpowers and the environment

Resource demand and environmental degradation

Superpower demand for resources leads to **environmental degradation**. China and the USA have the highest demand for most key commodities.

Environmental degradation can vary over a range of scales:

- Location (spatial) – local, national and global
- Time (temporal) – short-term, medium-term and long-term
- Severity – low, medium and high.

Resource	Demand	Environmental degradation
Food	• Meet dietary demands of consumer society • Food security (affordable, accessible, regular, nutritional supply) for emerging superpowers with growing populations	• Intensive agriculture leads to soil erosion and contamination through overuse of fertilisers and pesticides • Monoculture decreases biodiversity • Deforestation to increase farmland leads to habitat loss
Fossil fuels	• Meet energy needs through generation of electricity and fuel supply (for economic activity, households, transport)	• Increase in carbon dioxide emissions contributing to global warming • Increase in particulate matter leading to increase in respiratory disease
Minerals	• Raw materials for construction and manufacturing industries	• Open-cast mining scars landscape • Toxic chemicals released during mining processes
Water	• Water security (access to safe, affordable water for population) • Agriculture • Manufacturing • Energy	• Groundwater depletion through increased water demand • Water pollution through chemical runoff of waste products of agriculture and industry
Land	• Space for housing and economic activities in urban and rural areas	• Habitat and species loss due to urban sprawl • Encroachment on animal territories leads to conflicts between wildlife and residents

Carbon emissions

- Between them, superpowers' **carbon emissions** are significantly higher than other countries.
- China, the United States, the EU27, India, Russia and Japan were the world's largest CO_2 emitters in 2021.
- These countries have:
  — 49.2% of global population
  — 62.4% of global GDP
  — 66.4% of global fossil fuel consumption
  — 67.8% of global fossil CO_2 emissions.

Global agreements

Global agreements on environmental issues include:

- the United Nations Framework Convention on Climate Change (UNFCCC) aims to combat dangerous human interference with the climate
- the Kyoto Protocol (adopted in 1997, came into force in 2005) provided mechanisms to support reducing GHG emissions by 2012; the Doha amendment extended this commitment 2013–2020
- the Paris Agreement (2015) has the aim of keeping global temperature no more than 2°C (preferably 1.5°C) above pre-industrial levels

Global action on the environment

Different superpowers have different attitudes towards global agreements on environmental issues

Agreement	USA	China
UNFCCC	✓	✓
Kyoto Protocol to reduce emissions by 2012	✗	✓ without legally binding targets
Doha Amendment to Kyoto Protocol to reduce emissions 2013–2020	✗	✓
Paris Agreement to keep global temperature no more than 2°C (preferably 1.5°C) above pre-industrial levels	✓ (withdrew 2020 and rejoined 2021)	✓

USA
• The USA signed the Kyoto Protocol in 1997 but did not ratify it due to concerns that emerging superpowers China and India did not have legally binding emission reduction targets, and its economy would be disadvantaged as a result.
• Although the USA signed the Paris Agreement (2015), Republican President Donald Trump withdrew from the agreement in 2020 asserting it was unfair towards the USA. Democrat President Joe Biden re-signed the agreement in 2021.

China
• As a key player, it is essential that China is included in any global agreements on the environment.
• It signed the Kyoto Protocol (without legally binding limits) and the Paris Agreement.
• China's progress will be key to achieving the Paris Agreement targets.

Growth of middle-class consumption

- In 2023, the 'middle class' is the highest-spending part of the global population, responsible for more than 66% of the world's spending.
- By 2030, it is predicted that over 50% of the world's population (more than 4.2 billion people) will be middle class.
- The middle classes have higher disposable income than those in the working class.
- In the USA, the middle class is defined as households with between two-thirds more and double the median household income.
- In China, over 500 million people were added to the middle-/upper middle-income group between 2011 and 2019.
- Increases in the middle-class population will lead to future growth in **middle-class consumption**, particularly in emerging superpowers, leading to availability and cost issues as well as impacts on the environment.

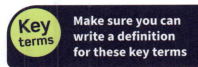 **Key terms** Make sure you can write a definition for these key terms

carbon emissions   environmental degradation
middle-class consumption   rare earth minerals   staple grains

9 Superpowers

9.6 Superpowers and the environment

Key resources

Rare earths (minerals)	• Electronic components used in high-technology devices rely on **rare earth minerals** (e.g. scandium). • As demand rises, superpowers invest in mining companies and technology, often in other countries. • Supplies of these minerals are finite, so there is competition to obtain them.
Oil	• Demand for energy including fuel for transportation (petrol-based vehicles), will rise. • Increased competition can lead to increased oil prices on the global market. • Any increase in the market share of electric vehicles may reduce this demand.
Staple grains	• Increased populations will increase consumption, leading to a rise in demand for **staple grains**. • Rice, maize, and wheat are staple foods for more than half the global population. • Demand for meat and processed foods increases as diets change and the ability to buy convenience foods rises.
Water	• Increased consumption puts pressure on freshwater resources. • Agriculture, industry, and households compete for limited supplies. • Water scarcity and conflict are key future issues.

Impacts on the physical environment

- Increased pollution (e.g. water, air, noise, land) and emissions reduce environmental quality and contribute to climate change.

- Deforestation: clearing of natural habitats (e.g. tropical rainforests) especially for agriculture and urban development. It leads to habitat loss, reduced biodiversity, loss of carbon sequestration potential, and ecosystem disturbance.

- Waste: increased waste (e.g. electronics, plastics) as demand and consumption of consumer products increases. Waste may take years to biodegrade in landfill sites and leak toxic substances into the ground.

- Resource depletion: finite resources are depleted when extracted and used. Unsustainable fishing practices reduce stocks.

▲ *Figure 1* *Deforestation of a rainforest to make way for palm oil and rubber plantations*

Increase in sustainable practices by governments and society.

Innovation and technology to find alternative materials and increase efficiency and sustainability.

Possible solutions

International cooperation: resources can cross borders, so agreements are needed to manage shared resources, tackle pollution, and mitigate the impacts of climate change.

REVISION TIP

Make a memory map to show how the demand for resources by superpowers is causing negative environmental effects.

RETRIEVAL

Learn the answers to the questions below, then cover the answers with a piece of paper and write as many as you can. Check and repeat.

Questions | Answers

#	Questions		Answers
1	What is an IGO?	Put paper here	Intergovernmental organisation
2	What is the IMF?		International Monetary Fund
3	What does the WTO do?		Promotes free global trade through a system of rules and regulations
4	Which IGO is a not-for-profit organisation linking the public and private sectors?	Put paper here	World Economic Forum (WEF)
5	What is a state-owned TNC?		A transnational corporation that is wholly or mostly owned by a government
6	What is an alliance?	Put paper here	A group of countries that formally work together on common interests
7	Name three types of alliances between countries.		Military, economic, environmental
8	What type of alliance is NATO?		Military
9	Name the IGO with 193 members cooperating on global issues such as peace, security, and sustainable development.	Put paper here	United Nations (UN)
10	Identify the five key resources in increasing demand by superpowers.		Food, fossil fuels, minerals, water, land

Previous questions

Now go back and use these questions to check your knowledge of previous topics.

Questions | Answers

#	Questions		Answers
1	What is a hyperpower?	Put paper here	A country that has all the characteristics of a superpower but remains unchallenged
2	What is an emerging power?		A country that has the potential to be a superpower in the future; it has significant economic power and is increasing its political, military, and cultural influence around the world
3	What characteristics can be used to define superpowers?	Put paper here	Economic, political, military, cultural, demographic, access to natural resources
4	What is hard power?		The use of military or economic force to gain influence
5	Name Rostow's five-stage model of economic development.		Modernisation Theory

9 Superpowers

9.7 Contested global influences

Tensions over resources

Superpower dominance relies on access to physical resources such as oil, gas, and minerals. Tensions exist between countries that disagree over their right to obtain these resources. This is due to border and territorial disputes, particularly when they impact the extent of a country's exclusive economic zone (EEZ). An EEZ is an area of 200 nautical miles from a country's coastline within which a nation controls sea and seabed resources. The UN settles disputes between countries over EEZs.

> **REVISION TIP**
>
> Draw a sketch map of a region where the exploitation of physical resources is disputed between countries. Annotate it with details to show why there is a dispute.

Disputed territorial claims in the Arctic

- Many countries claim parts of the seabed in the Arctic Ocean due to the significant resource reserves to be found there, including oil and gas, rare earth metals, precious metals and ores, and construction minerals.

- Shrinking sea ice cover due to warming Arctic temperatures is increasing shipping access to the region for more weeks of the year and making test drilling for oil, gas, and minerals less costly.

- Russia and NATO allies have increased military presence in the region, including military bases, warships, and military exercises. Both have nuclear weapons.

- China also has interests in the region. Warmer temperatures would open up more ice-free routes that remain navigable for more of the year. The Polar Silk Road would cut the distance from Shanghai to Germany by 5000 km.

- The eight Arctic state members of the Arctic Council (an IGO) promote cooperation but do not impose legislation. The UN acts as a broker when countries are in dispute.

Intellectual property: a source of tension

Intellectual property (IP) is protected in law through patents, copyrights, trademarks, trade secrets, geographical indications, and industrial designs. The UN's World Intellectual Property Organisation (WIPO) operates the global system for IP. Without legal protection, innovation and research can be stolen and copied, including through **counterfeiting**.

> **REVISION TIP**
>
> You need to know what threatens intellectual property rights and the negative effects of this.

Benefits of IP system	Criticisms of IP system
☑ Encourages investment in innovation	☒ TNCs have more money to protect their IP
☑ Stimulates economic growth	☒ Creates monopolies and market power – anticompetition, increased prices, decreased consumer choice
☑ Protects inventors and creators	
☑ Boosts consumer confidence through trusted, safe products	☒ High cost of products (e.g. patented medicines) can restrict access for consumers
☑ Important for global trade	☒ Users of IP protected products may have to pay fees or buy a licence; this creates a barrier for LICs
☑ Protects products made in a unique location from imitation	☒ Complex, costly legal battles over IP rights

Countries that do not tackle counterfeiting may damage their global reputation:

- Economically, other countries are less likely to enter into trade agreements with countries weak on dealing with counterfeiting.

- Politically: if a country does not uphold IP laws, it indicates they may have a less robust attitude to other international legislation.

> The USA and China have tense IP relations, particularly linked to technology and trade. In 2022, for example, the USA added E-commerce sites operated by Chinese firms Alibaba and Tencent to its list of companies it believed to be involved in trading counterfeit goods.

Political spheres of influence

A **political sphere of influence** is an area that a country has power over without formal territorial control. Spheres of influence can include both terrestrial and maritime areas. Disputes arise when different countries contest spheres of influence.

REVISION TIP

Create a memory map to show the reasons why some countries disagree about territory, causing conflict.

Border disputes: e.g. unresolved historical disagreements resulting from colonialism, imperialism, or previous conflict

Territorial claims: e.g. Falkland Islands colonised as British Overseas Territory but claimed by Argentina

Securing existing, newly discovered, or potential resources: e.g. oil and gas

Reasons for disputes

Expansion of influence in regions with shared ethnic, linguistic, or cultural ties

Competition for trade routes

Proxy conflicts: superpowers support political groups in smaller countries to extend their sphere of influence without direct confrontation

Ideological differences: e.g. capitalist versus communist ideologies in Europe during the Cold War

South and East China Seas

The South and East China Seas area is important for China.

Key for China's national security, including securing the reunification of Taiwan with mainland China. Many neighbouring countries are claiming parts of the region as their own sovereign territory.

Claims on key natural resources, e.g. fishing grounds, oil and gas reserves.

Increasing its regional influence both militarily and economically.

Countering US influence in the Asia Pacific region.

Key
- – – China's claim line
- - - - UNCLOS 200 nautical mile EEZ
- ○ Disputed islands

CHINA

TAIWAN

South China Sea

VIETNAM

MALAYSIA

MALAYSIA

PHILIPPINES

INDONESIA

500 km
200 miles

China and Japan are involved in resource-based territorial disputes in the East China Sea.

Area includes key trade routes.

The US has not taken sides in the disputes but has increased its military operations in the region.

China has made claims over small and largely uninhabited islands and maritime zones in the South China Sea. China has backed up this expansion by building artificial islands and increasing its military presence (military bases and patrols).

Neighbouring countries competing for claims include Vietnam, the Philippines, Malaysia, Taiwan, and Brunei.

▲ **Figure 1** *South and East China Seas*

9 Superpowers

9.7 Contested global influences

Western Russia/Eastern Europe

Occupation and annexation of other regions can lead to war.

Collapse of the USSR creates new independent states in Eastern Europe.

Russia annexes the Crimean Peninsula in southern Ukraine. Russia is removed from the G8 and sanctions are imposed. Pro-Russian activists gain control of government buildings in the Donbas region.

| 1991 | 2004 | 2014 | 2022 |

Estonia, Latvia, and Lithuania (which border Russia) join the EU and NATO.

Russia launches its invasion of Ukraine.

Russian invasion of Ukraine

- Russia and Ukraine have deep historical ties and were both part of the Soviet Union.
- Russia sees Ukraine as strategically important due to its size, location, and Black Sea access. Crimea provides a warm-water port for the Russian navy.
- Russia was concerned about EU and NATO expansion. Ukraine aspires to join both.
- Russia seeks to expand its buffer zone to increase its security. Ukraine borders Belarus, a pro-Russian state. The 2014 Ukraine revolution ousted the pro-Russian president and installed a pro 'Western' government.

- Regions in Eastern Ukraine have a significant ethnically Russian population. Russia claimed the need to protect their rights and linguistic and cultural ties.
- Ukraine is a market for Russian goods and a key route for transporting Russian gas to Western Europe. Controlling this allows Russia to put pressure on the 'West'.
- Russian President Vladimir Putin has promoted nationalist and patriotic feeling to maintain his domestic support and power.

Impact of conflict on people and the environment

People	Built and natural environment
• Death, physical/mental injury and increased disease	• Destruction of key infrastructure
• Displacement – migration to neighbouring or supporting countries	• Pollution from explosives, unexploded mines, hazardous chemicals
• Family separation	• Habitat destruction including deforestation
• Loss of property and livelihoods	• Over-exploitation/illegal exploitation of resources
• Reduced access to essential resources (food, water)	• Disruption of agriculture – food shortages, price increases
• Disruption of education and healthcare systems	• Soil erosion
• Human rights violations	• Impact on marine ecosystems
• Need for humanitarian aid, short- and long-term	
• Social tensions between ethnic groups	
• Disruption of trade networks	

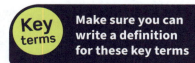 **Make sure you can write a definition for these key terms**

counterfeiting   intellectual property (IP)
political sphere of influence

9.8 Superpowers and developing nations

Economic ties between emerging powers and the developing world

Benefits for emerging powers	Benefits for developing world
• Gain valuable in-demand natural resources • Gain access to new market for goods and services • Make geopolitical and strategic ties	• Gain financial resources, aid, and FDI • Gain expertise in large infrastructure projects • Obtain technology • Gain access to large markets • Lessen dependence on other trade partners, e.g. ex-colonists

China and Africa

Trade

The value of China–Africa trade has been increasing since 2000.

Trade is more favourable for China – the value of goods China exports to Africa is more than the value of goods Africa exports to China. Raw materials and crops China buys from Africa are cheaper than the manufactured goods African countries buy from China.

African countries have become part of China's BRI, a long-term project to expand China's economic and political sphere of influence through investment in projects (e.g. infrastructure) both in China and abroad.

FDI

Since 2013, Chinese FDI flows into Africa have been higher than those from the USA.

FDI can be used for energy and transport infrastructure projects (e.g. HEP, road, and rail).

Aid

Chinese development assistance is given to African projects linked to economic and social development as well as political and economic stability. In the early 2010s, Chinese loans to African countries were worth more than those from the USA. Chinese assistance is increasingly likely to be loans, charging interest over a certain time period, than grants.

Environmental impacts in Africa

• Land degradation through open-cast mining.
• Loss of habitats such as tropical rainforest and savannah through land development.
• Contamination of water and land through toxic waste and oil spills.

Opportunities and challenges

Opportunities of the China–Africa economic relationship	Challenges of the China–Africa economic relationship
• FDI helps Chinese companies to grow their business. • FDI helps African countries to make progress with development goals. • FDI creates employment opportunities for local people. • Aid projects benefit local people and help countries progress towards sustainable development goals (SDGs). • Port infrastructure aids economic growth and provides bases for military security. • Opening up of landlocked countries to the global economy. • Progress in transnational projects with weak governments.	• Protests by local people towards Chinese projects. • 'Sinoisation' – neo-colonialism by China leading to possible exploitation of African resources for Chinese gain. • Rising debt issues. • Environmental challenges (see above). • Cancellation and delays of projects by African governments. • Employment rights and safety issues. • Limited progress in transnational projects with strong tensions between countries. • Limited success with third-country partners.

⚙ KNOWLEDGE

9 Superpowers

9.8 Superpowers and developing nations

The rise of China and India

China and India have increased their geopolitical and economic influence but also raise regional tensions.

China		India
South and East China Sea Taiwan Hong Kong Tibet Indian border dispute	Complex relationship with North Korea Island disputes with Japan Tensions with Australia	Disputes with Pakistan over Kashmir Chinese border dispute Nepal border dispute Illegal migration from Bangladesh Fishing disputes with Sri Lanka

Tensions in the Middle East

The geopolitics of the Middle East are complex due to historical, cultural, and economic factors. The challenge for superpowers is to maintain political and economic ties whilst negotiating regional tensions and human rights issues.

Country	Allies	Tensions
Israel	USA and its allies	Tensions over Palestinian State
Iraq	USA and its allies	Tensions following Iraq War
Syria	Russia and Iran	Ongoing civil war
Iran	Syria	Sanctions imposed by USA and allies because of nuclear weapon development
Arabian Peninsula (Saudi Arabia, UAE, Qatar, Kuwait)	USA and its allies	Between Saudi Arabia and Yemen
Türkiye	NATO member	Tensions with Kurds who want to establish their own nation state

Reasons for instability in the Middle East

- Historical conflicts and colonial legacy lead to border disputes.
- Diverse religious and sectarian groups: Sunni and Shia Muslims, Christians, Jews, and other communities. Regional majority/minority relations can lead to tensions.
- 65% of known global oil reserves and 38% of known global gas reserves make the region important for the economic development of other countries and a source of global competition for resources.
- Hot dry climate leads to pressure on freshwater reserves, including aquifers and transboundary water sources.
- Limited agricultural land and increased reliance on food imports.

- Unstable or democratically weak systems of government and lack of political freedom for some citizens.
- Rising youth populations seeking new opportunities and improved human rights have led to uprisings and pro-democracy protests.
- Economic inequality between different groups.
- Rise of terrorist and extremist groups not aligned to governments.
- High migrant worker populations in some states, leading to human rights issues.
- Management of ongoing refugee crises (e.g. Türkiye–Syria border).

9.9 Challenges facing superpowers

Economic restructuring and unemployment

- As the world's economic centre shifts towards the East, 'western' superpowers have faced many economic issues.
- **Economic restructuring** (shifting the economy so a different sector is dominant) in the USA and EU has led to a decrease in the number of jobs in manufacturing and an increase in tertiary and quaternary employment.
- TNCs may favour offshoring and outsourcing of manufacturing and services to regions of the world where labour is less expensive (Asia and Africa).
- Job opportunities in the service sector may also decline due to the growth of online services and artificial intelligence (AI). Superpowers may have to invest in workforce education and training and develop different work patterns.

Economic issues linked to population

- Superpowers may compete for international migrants, particularly in high-skill or skill-shortage sectors.
- Host countries will also have to manage integration of migrants.
- Emerging superpowers have youthful populations and will benefit from the demographic dividend (growth in an economy due to the change of a country's age structure).

EU	USA
The proportion of the population that is economically active is decreasing leading to labour shortages and less tax revenue. The proportion of the population that is retired (65+) will increase causing pressure on health and social care services and the cost of state pensions. Increased migration will be required to provide workers in key professions.	The US population is increasing. The US population will age. Economic migrants will be relied on as a labour supply in shortage sectors.

Debt and financial issues

- Global recessions have slowed the economic growth of superpowers. The 2008 global financial crisis, which arose largely from the mortgage-lending market in the USA and Europe, has increased the debt level of many governments.
- Austerity measures introduced following the financial crisis and subsequent recession have reduced public spending.
- Some Eurozone members (e.g. Greece) required European Central Bank intervention to stabilise their economies, following increases in public debt and debt servicing.
- EU Regional Development Funds were invested in countries joining the EU from 2004 to reduce regional inequalities.
- The response to the global COVID-19 pandemic has had a high economic cost, including subsidies for workers and businesses affected by the pandemic, the cost of vaccine research, and increased healthcare costs.
- The EU has experienced a cost-of-living crisis (2021 onwards) due to increased inflation of the price of goods and services, and rising housing costs.

Social costs

- Economic restructuring and the resulting unemployment have led to a loss of social cohesion and an increase in mental health problems in many formerly industrialised regions.
- Many people have been forced to migrate from these regions in order to find work.

9 Superpowers

9.9 Challenges facing superpowers

Maintaining global power

It is generally assumed that superpowers must spend money on **military power** to preserve national security and maintain their global influence. However, it is expensive.

In light of economic issues and rising healthcare costs, it is often debated whether high levels of military spending should continue.

Naval and air power	Nuclear weapons	Intelligence services
• Expensive research and development are required to ensure that ships and aircraft are improved and maintained. • Rapid-response aircraft are becoming increasingly important.	• Nuclear weapons are designed as a deterrent, so many argue that the defence industry (primarily in the USA) is the main beneficiary of continued high investment in nuclear strength. • Emerging powers are investing in nuclear weapons. In 2016, the UK government voted to renew the UK's Trident nuclear deterrent with an updated version.	• High-profile terrorist attacks have led to increased government spending on intelligence. • Intelligence work is high-tech but labour intensive, so the costs are high.

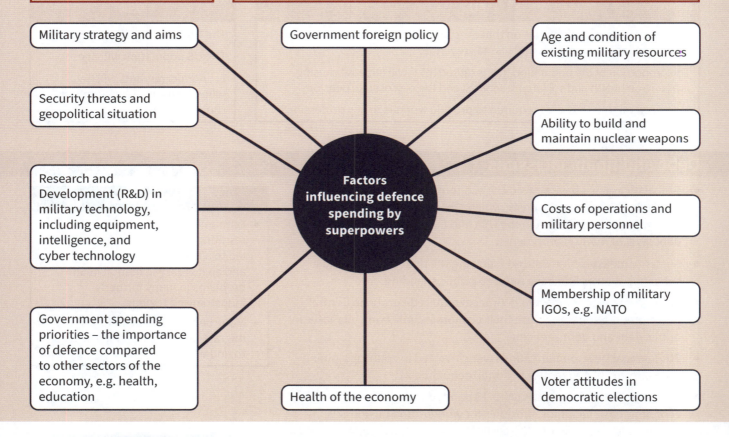

Factors influencing defence spending by superpowers:
- Military strategy and aims
- Government foreign policy
- Age and condition of existing military resources
- Security threats and geopolitical situation
- Ability to build and maintain nuclear weapons
- Research and Development (R&D) in military technology, including equipment, intelligence, and cyber technology
- Costs of operations and military personnel
- Government spending priorities – the importance of defence compared to other sectors of the economy, e.g. health, education
- Membership of military IGOs, e.g. NATO
- Health of the economy
- Voter attitudes in democratic elections

 Key terms Make sure you can write a definition for these key terms

economic restructuring   intelligence services   military power

Space exploration

The USA remains the dominant force in space exploration through the National Aeronautics and Space Administration (NASA). In the light of economic issues such as rising healthcare costs, some question whether high levels of spending on space exploration should continue. However, NASA plays a key role in maintaining superpower status for the USA through contributions to:

- National security and planetary defence, e.g. tracking space debris and asteroids which could impact Earth
- Scientific exploration
- Technological developments which are now essential to society, e.g. satellite communication, GPS

- International collaboration
- Global prestige
- Promoting STEM (Science, Technology, Engineering and Maths) education and research

Space exploration in emerging economies

Space budgets may be under threat in many 'Western' countries, but India and China are often able to launch space missions more cheaply:

- The Indian Space Research Organisation (ISRO) has launched satellites, a Mars orbiter, a lunar south pole lander and Sun probe.
- The China National Space Administration (CNSA) has landed a spacecraft on the far side of the moon.

> **REVISION TIP**
> Write down on flash cards the reasons why superpowers are questioning the costs of military power and space exploration.

The future balance of global power

The balance of global power in 2030 and 2050 is uncertain. There are a range of possible outcomes:

- Uni-polar – one superpower (USA) is the dominant global player (hegemon).
- Bi-polar – two superpowers (USA and China) are dominant global players with different ideologies and competing spheres of influence.
- Multi-polar – more than two countries or regions play significant roles in the global power structure (USA, China, EU, Russia, and India). Conflict resolution and cooperation becomes more complex.

Factors affecting the future global balance of power

- Global population growth is slowing down.
- Ageing population in USA, Europe, Japan, China leading to increased healthcare and pension costs, and labour shortages.
- Demographic dividend for India: economic benefit due to increase in proportion of working population.
- Rise of Asia: by 2040, Asia is likely to contribute more than 50% of global GDP and account for 40% of global consumption.
- New and existing conflicts and the threat of terrorism.
- Technological development: AI, quantum computing, renewable energy, biotechnology.
- Response to environmental challenges.
- Rise of power of players other than sovereign states (e.g. TNCs, military groups).
- Uncertainties in the global economy.
- Changing trade relations, including the influence of trade blocs and FDI.
- Digital governance, including data privacy and cybersecurity.
- Relationships with IGOs, including those linked to global governance (e.g. UN).

> **REVISION TIP**
> Try to explain to a classmate what you think the balance of global power will be in 2050, and why you think this.

RETRIEVAL

Learn the answers to the questions below, then cover the answers with a piece of paper and write as many as you can. Check and repeat.

Questions

1 What is an exclusive economic zone (EEZ)?

2 What is intellectual property (IP)?

3 What is counterfeiting?

4 What is a political sphere of influence?

5 What are the three environmental impacts on Africa of China's increasingly close relationship with the continent of Africa?

6 Who are Syria's allies?

7 What is economic restructuring?

8 What is the demographic dividend?

Answers

Put paper here

An area up to 200 nautical miles from a country's coastline in which a nation controls sea and seabed resources

The protection of ideas and inventions through trademarks, patents, copyright, and other mechanisms

The fraudulent imitation of a product

An area that a country has power over, often dominating that region without formal territorial control

Land degradation through open-cast mining; loss of habitats such as tropical rainforest and savannah through land development; contamination of water and land through toxic waste and oil spills

Russia and Iran

Shifting the economy so a different sector is dominant

Growth in an economy due to the change of a country's age structure

Previous questions

Now go back and use these questions to check your knowledge of previous topics.

Questions

1 What is an IGO?

2 What is a state-owned TNC?

3 What is an alliance?

4 Identify the main reasons for increasing resource demand in superpowers.

5 Name the European economic and political organisation with free movement of people, goods, services, and capital between member states.

Answers

Put paper here

Intergovernmental organisation

A transnational corporation that is wholly or mostly owned by a government

A group of countries that formally work together on common interests

Rising populations; high income populations with high consumer demands; economic sectors need energy and raw materials for production; transport

European Union (EU)

 **PRACTICE**

Exam-style questions

1   Explain **one** way in which **one** named geo-strategic location theory
    links territory to political power.   **(4)**

2   Explain **one** reason why direct colonial control was used in the
    British Empire.   **(4)**

3   Complete **Figure 1** by filling in boxes A–D in the final two rows.   **(4)**

EXAM TIP

Some of these 4-mark questions will assess your quantitative skills. 'Complete' requires you to add detail to a graph or table that has been provided using the data that is also provided.

	Brazil	China	India	Russia	South Africa
Population (million)	215	1 412	1 417	144	59
Rank	3	2	1	4	5
Weighted rank (Rank × 1)	3	2	1	4	5
GNI per capita ($)	17,270	21,250	8230	35,540	15,590
Rank*	3	2	5	1	4
Weighted rank (Rank × 2)	6	4	10	2	8
Individuals using the internet (% of population)	79	76	46	90	72
Rank*	2	3	5	1	4
Weighted rank (Rank × 1)	2	3	5	1	4
Military expenditure ($ billion)	20	291	81	86	3
Rank*	4	1	3	2	5
Weighted Rank (Rank × 1)	4	1	3	2	5
Mortality rate attributed to household and ambient air pollution (per 100 000)	28	95	139	67	74
Rank*	1	4	5	2	3
Weighted rank (rank × 0.5)	0.5	**A**	**B**	1	1.5
Overall score	**C**	12	21.5	10	**D**

*Ranks are from 1–5, where 1 is the highest and 5 is the lowest.

▲ *Figure 1 A table comparing BRICS using selected data with weighted ranks (Data from 2019–2022) (Data source: World Bank)*

4   Explain **one** way in which powerful nations exert cultural influence.   **(4)**

EXAM TIP

For a 4-mark 'Explain' question, your answer should provide a reasoned explanation of how or why something occurs, demonstrating your understanding of the topic through your justification and/or examples.

5   Explain **one** reason why some countries choose to join military alliances.   **(4)**

6   Explain **one** reason why superpower resource demands can cause
    environmental degradation.   **(4)**

7   Explain **one** way in which the global system of intellectual property rights
    can be threatened.   **(4)**

Exam-style questions

8   Explain **one** reason why tensions in the Middle East present an ongoing
     challenge to superpowers.   (4)

9   Explain **one** reason why space exploration is questioned in some existing
     superpowers.   (4)

10   Assess the view that soft power is as important as hard power in
     maintaining superpower status.   (12)

11   Assess the view that a unipolar world brings more geopolitical stability
     and poses fewer risks than bi- or multipolar worlds.   (12)

12   Assess the relative usefulness of development theories to help explain
     changing patterns of power.   (12)

13   Assess the relative importance of superpowers in influencing IGOs. (12)

14   Assess the relative success of the UN in maintaining global
     geopolitical stability.   (12)

15   Assess the relative importance of the future growth in middle-class
     consumption within emerging superpowers on the physical environment. (12)

16   Assess the view that economic factors contribute more than political
     factors to tensions arising over the acquisition of physical resources. (12)

17   Assess the view that more opportunities than challenges are created through
     economic ties between emerging powers and the developing world.   (12)

18   Assess the view that it will be economic factors will affect future power
     structures in a 21st century world.   (12)

> **EXAM TIP**
>
> The command word 'Assess' requires you to use evidence to determine the relative significance of something. Your answer needs to be balanced, considering all relevant factors and identifying which is the most important factor, and why.

 **KNOWLEDGE**

10 Health, human rights and intervention

10.1 Human development

Measuring human development

- The dominant models of development are 'western' in origin and are contested.
- It can be argued that **GDP** is a good measure of development, as economic development drives other improvements within a country.
- However, GDP may not be the best indicator, as the concept of development is complex and multi-stranded.
- A high GDP does not always mean that a country is socially, environmentally, culturally, and politically developed.

> **REVISION TIP**
>
> You need to know the difference between GDP and GDP per capita and understand why this is a contested method of assessing development in a country.

GDP as a measure of development

- In 2021, the USA had the highest GDP (23 trillion US$) and Tuvalu had the lowest (63 million US$).
- **GDP per capita (PPP)** measures the amount of wealth a country generates within its borders, divided by the population size.
- PPP means that adjustments have been made so data can be compared between countries with different currencies.
- In 2021, Luxembourg has the highest GDP per capita, and Burundi had the lowest.

Happy Planet Index

- The Happy Planet Index is an indicator that contests the view that GDP is the strongest measure of development.
- It considers wellbeing, life expectancy and ecological footprint.
- In 2019, Costa Rica scored highest, and Qatar scored lowest. In 2021, Qatar had the fourth highest GDP per capita.
- High levels of inequality or corruption may mean that levels of human contentment and quality of life are not always the highest in the wealthiest nations.

Alternatives to dominant Western models

Bolivia under Evo Morales

Evo Morales was Bolivia's president between 2006 and 2019. He was a socialist who taxed TNCs, increased the minimum wage, and renationalised the oil and gas industries. Bolivia had high levels of poverty and inflation but, over this period, the country experienced a fall in poverty, a decline in economic inequality, and steady economic growth. However, Bolivia still has the lowest GDP per capita of the 11 South American countries.

> **REVISION TIP**
>
> You need to know about how dominant models of human development are contested in one place context.

Goals for development

- Some people see improvements in environmental quality, health, life expectancy and human rights as more significant goals for development than economic wealth.
- Swedish academic Hans Rosling believed that improving education, women's rights, and access to healthcare, including contraception, is the key to development.
- Rosling's data showed that, as health in developing countries improves, infant mortality declines, birth rates fall, and people are lifted out of poverty.
- Governments should invest wealth from economic growth in health and education in order to address poverty.
- Rosling also identified the environment as a threat to development: climate change will affect fertility, health, death rates, and migration.

10 Health, human rights and intervention

10.1 Human development

Education

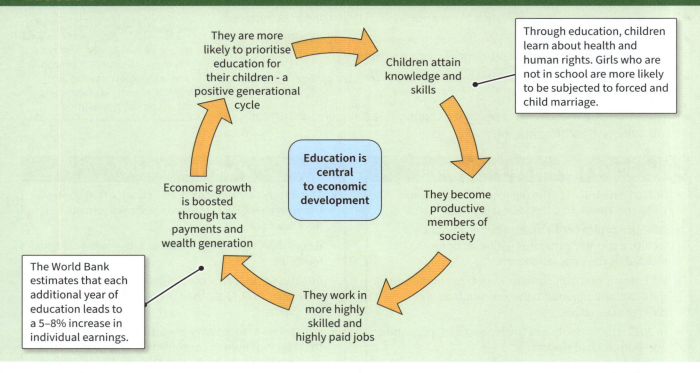

They are more likely to prioritise education for their children - a positive generational cycle

Children attain knowledge and skills

Through education, children learn about health and human rights. Girls who are not in school are more likely to be subjected to forced and child marriage.

Education is central to economic development

Economic growth is boosted through tax payments and wealth generation

They become productive members of society

The World Bank estimates that each additional year of education leads to a 5–8% increase in individual earnings.

They work in more highly skilled and highly paid jobs

Different views on education

The view that education is important is not universally shared. 16% of countries still do not have the right to education in their laws. In some countries, females face barriers to education. UNESCO estimates that, globally, 129 million females under 18 are out of education.

This could be due to:

- families in poverty prioritising male education
- girls being subject to child marriage and dropping out of school
- girls being expected to work in the household and in childcare instead
- inadequate toilet and sanitary facilities
- lack of female teachers and gender bias in the curriculum local laws – in Afghanistan, girls over the age of 11 are prohibited from going to secondary school.

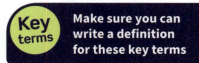 **Key terms** Make sure you can write a definition for these key terms

GDP   GDP per capita (PPP)   Happy Planet Index

10.2 Health and life expectancy

Variations in health and life expectancy in the developing world

In the developing world, there are lower levels of health and lower life expectancies.

- Most deaths are due to communicable diseases.
- Healthcare is more likely to be of poorer quality and inaccessible, especially in rural areas.
- The rates of **maternal mortality**, **child mortality** and life expectancy in a country reflect its healthcare services, the accessibility of medical care, the number of trained medical personnel and advancements in treatment.
- Food insecurity can lead to undernourishment and malnourishment.
- People, particularly children, who are malnourished are more likely to die from hunger and hunger-related illnesses. Rates of malnourishment are highest in Nigeria and India.
- Lack of access to potable water and **sanitation** increases the spread of waterborne diseases such as cholera and typhoid.
- Diarrhoea is the second leading cause of death in children. It results from contaminated food and water sources.
- Many developing countries are situated in the tropical regions of the world, where climatic conditions are suitable for vectors which spread diseases such as malaria and dengue fever.

> Although global health and life expectancy are increasing, there are considerable variations.

↓

> Variations are explained by differential access to basic needs such as:
> - food
> - water supply
> - sanitation.

↓

> These impact heavily on levels of infant and maternal mortality.

Life expectancy in the developing and developed world
In 2019: • Japan had the highest life expectancy at birth (85 years) • the Central African Republic had the lowest (53 years).

10 Health, human rights and intervention

10.2 Health and life expectancy

Variations in health and life expectancy in the devleoped world

Health and life expectancy in HICs is higher. However, there are variations within populations in the developed world.

- Most deaths are due to non-communicable diseases, sometimes caused by lifestyle factors. Rates of smoking, alcohol consumption and overnutrition are higher in some developed countries.

- There are higher rates of some cancers and this reduces life expectancy.

- Higher levels of **deprivation** will impact someone's ability to buy healthy, nutritious food and live in high-quality housing, both of which will affect health outcomes.

- In some developed countries, healthcare may be expensive or it may take time to receive treatment.

- Some countries, such as Norway provide universal health coverage for their population and invest heavily in healthcare services.

Variations within countries

- There are significant variations in health and life expectancy within countries.

- Variations in income will lead to inequalities in access to resources that enable a healthy lifestyle, and higher rates of engagement with risk factors, such as smoking.

UK	Indigenous peoples in Australia
In the UK, there are healthcare inequalities between some groups and communities. Women living in the most-deprived areas of England have a life expectancy 8 years lower than those living in the least-deprived areas. Those living in areas with higher deprivation are more likely to lack access to healthcare services. This is due to: • lack of access to transport • language barriers • literacy levels • misinformation and previous negative experiences.	Variations in health and life expectancy within countries can be related to ethnic variations. In Australia, the burden of disease (impact of health problems) is 2.3 times higher for Indigenous peoples than for non-Indigenous peoples, and life expectancy is approximately 17 years lower. This is due to: • lack of accessible healthcare services • high cost • poorer quality infrastructure • lack of culturally appropriate services.

> **REVISION TIP**
>
> You need to know about variations in health and life expectancy in one place context and in relation to ethnic variations in another place context.

Key terms Make sure you can write a definition for these key term

| child mortality | deprivation | maternal mortality | sanitation |

10.3 The role of governments and IGOs

The role of governments

Decisions made by governments can influence the relationship between economic and social development. Countries with higher expenditure on healthcare tend to have higher life expectancies and lower levels of child mortality.

Welfare states prioritise spending to improve the economic and social wellbeing of citizens, for example through maternity benefits, pensions and health services. Norway has one of the highest percentages of social spending and scores the highest out of all OECD countries in the Happy Planet Index.

Some countries have low levels of spending on health and education. Spending on education in Nigeria is one of the lowest in the world. The literacy rate was 62% in 2018.

The role of IGOs

World Bank
A financial institution that provides loans and funding to developing countries, NGOs and environmental organisations.

International Monetary Fund (IMF)
The financial agency of the United Nations which gives loans to countries in financial difficulty and advises countries on policies to improve economic stability.

World Trade Organisation
An organisation that facilitates trade between countries by negotiating agreements.

Traditionally, the dominant **IGOs** have promoted **neo-liberal** views of development, which encouraged:

- the adoption of free trade by removing trade barriers
- privatising services and industries that had been owned by the state
- deregulating financial markets by removing restrictions on businesses to make it easier for them to trade.

1980s	1996	Today
Structural Adjustment Programmes (SAPs) involved the World Bank or IMF lending a developing country money to boost economic growth and reduce poverty. Countries also had to implement neo-liberal policies. SAPs were criticised for opening developing counties up to exploitation and increasing their debt.	The IMF and World Bank launched the Heavily Indebted Poor Countries Initiative, which cleared the unsustainable debts owed by 36 eligible developing countries, enabling them to spend more on health and education services instead of debt repayments.	There are programmes aimed at improving environmental quality, health, education and human rights such as: The World Bank's water and sanitation project has increased access to clean water for 560,000 people in Haiti's rural areas. IGOs also fund a project supporting Indonesia to improve the lives of coastal communities and increase resilience to sea level rise.

10 Health, human rights and intervention

10.3 The role of governments and IGOs

The UN's Millennium Development Goals (MDGs)

The United Nation's 8 **MDGs** were created between 2000 and 2015. They encouraged countries to work together to achieve the goals shown below:

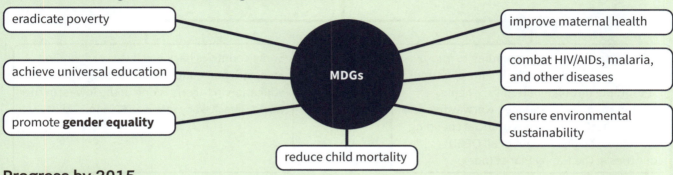

- eradicate poverty
- achieve universal education
- promote **gender equality**
- reduce child mortality

MDGs

- improve maternal health
- combat HIV/AIDs, malaria, and other diseases
- ensure environmental sustainability

Progress by 2015

☑ Extreme poverty had declined.	☒ Gender inequality persisted.
☑ Primary school enrolment rate had increased.	☒ Inequalities between the wealthiest and poorest areas still existed.
☑ More girls were in school.	☒ Global carbon dioxide emissions had increased.
☑ The under-five mortality rate had declined by more than half.	☒ Water scarcity had increased.
☑ Maternal mortality had declined.	☒ Numbers of refugees increased.
☑ Malaria deaths had fallen and treatments for disease had improved.	☒ 800 million people still lived in poverty.
☑ Access to drinking water had improved.	☒ Poverty rate fell by 66% in Southern Asia but only 28% in Sub-Saharan Africa.
☑ Official development assistance from developed countries had increased.	☒ The target to reduce hunger was met in some regions (e.g. South America and Mexico), but not others, (e.g. Sub-Saharan Africa), and increased in Western Asia.

The UN's Sustainable Development Goals (SDGs)

In 2015, as part of the UN's post-2015 development agenda, 17 **SDGs** were created:

1. No poverty
2. Zero hunger
3. Good health and well-being
4. Quality education
5. Gender equality
6. Clean water and sanitation
7. Affordable and clean energy
8. Decent work and economic growth
9. Industry, innovation and infrastructure
10. Reduced inequalities
11. Sustainable cities and communities
12. Responsible consumption and production
13. Climate action
14. Life below water
15. Life on land
16. Peace, justice and strong institutions
17. Partnerships for the goals

Key terms

Make sure you can write a definition for these key terms

gender equality   IGOs   MDGs   neo-liberal   SDGs   welfare states

RETRIEVAL

Learn the answers to the questions below, then cover the answers with a piece of paper and write as many as you can. Check and repeat.

Questions	Answers
1 What does GDP per capita (PPP) measure?	The amount of wealth a country generates within its borders, divided by the population size
2 What is the Happy Planet Index?	An indicator of development that takes into account wellbeing, life expectancy and ecological footprint
3 List two measures that Evo Morales introduced in Bolivia.	Two from: taxed TNCs / increased the minimum wage / renationalised the oil and gas industries
4 Give one reason why healthcare inequalities exist within developed countries.	One from: Lack of access to transport / language barriers / literacy levels / misinformation and poor previous experiences
5 Give one reason why females face barriers to education.	One from: priorities for male education / child marriage / household and childcare / inadequate toilet facilities / lack of female teachers / gender bias in the curriculum / prohibited by the country leaders
6 Name one IGO that promotes development.	One from: IMF / World Bank / UN / WTO
7 List two measures encouraged in a neo-liberal view of development.	Two from: the adoption of free trade by removing trade barriers / privatising services and industries that have been owned by the state / de-regulating financial markets by removing restrictions on businesses to make it easier for them to trade.
8 What are the UN's MDGs?	Millennium Development Goals
9 State three of the UN's MDGs.	Three from: eradicate poverty / achieve universal education / promote gender equality / reduce child mortality / improve maternal health / combat HIV/AIDs, malaria, and other diseases / ensure environmental sustainability / encourage countries to work together
10 What are the UN's SDGs?	Sustainable Development Goals
11 State three of the UN's SDGs.	Three from: No poverty / Zero hunger / Good health and well-being / Quality education / Gender equality / Clean water and sanitation / Affordable and clean energy / Decent work and economic growth / Industry, innovation and infrastructure / Reduced inequalities / Sustainable cities and communities / Responsible consumption and production / Climate action / Life below water / Life on land / Peace, justice and strong institutions / Partnerships for the goals

Put paper here

10 Health, human rights and intervention

10.4 Human rights laws and agreements

The Universal Declaration of Human Rights (UDHR)

- The **UDHR** is a set of statements, compiled by the UN in 1948, which outlines the rights that all people around the world should be entitled to without discrimination.

- It is not legally binding but has inspired international human rights law and treaties.

- In 1948, eight countries abstained from voting in favour of the UDHR. Signing and ratifying the UN's human rights conventions (legal agreements between countries and the UN) and treaties remains optional for governments.

The European Convention on Human Rights (ECHR)

- The **ECHR** was created in 1950 by the member states of the Council of Europe. All new EU member states are expected to ratify it.

- It created the European Court of Human Rights where cases relating to violations of the convention can be heard.

- The rights outlined in the ECHR were made part of the UK law through the Human Rights Act of 1998.

- The UK is still part of the Council of Europe and subject to the rights outlined in the ECHR.

- The ECHR can be controversial as some see it as an erosion of national **sovereignty** that takes decisions and judgements on violations of rights away from individual countries.

The Geneva Conventions

- The **Geneva Conventions** are four treaties which outline the rights of and protections for those engaged in warfare. They are a large part of international humanitarian law.

- They were created between 1864 and 1977 and adopted by 196 nations.

- They protect wounded soldiers and civilians, give them the right to medical care, prohibit torture, and ensure humane treatment of prisoners of war.

- They are used to bring those who commit war crimes to trial; however, few cases reach this stage and over 150 countries continue to engage in torture.

- The International Committee of the Red Cross promotes respect for these international humanitarian laws.

> **REVISION TIP**
>
> Make sure you how the UDHR, ECHR and Geneva Conventions are used to protect and promote human rights.

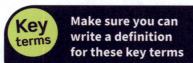 **Key terms** Make sure you can write a definition for these key terms

ECHR   Geneva Conventions   sovereignty   UDHR

10.5 Human rights variations between countries

Global variations in human rights

There are significant differences between countries in both their definitions and protection of human rights.

Some states frequently invoke human rights in international forums and debates, for example the UK.

UK
• The UK is on the UN Security Council, UN Human Rights Council, the Council of Europe, and has signed many of the UN's international human rights treaties.
• In 2017, at the 72nd Meeting of the UN General Assembly. The UK government issued a 'Call to Action' to end modern slavery. An independent assessment recognised in the UK as the country making the most progress towards the UN's SDG on modern slavery.
• Human rights in the UK are protected by the 1998 Human Rights Act, which is based upon the European Convention on Human Rights.

Other states prioritise economic development over human rights and defend this approach for example, Singapore.

Singapore
• Singapore is an advanced economy with one of the highest GDPs in the world.
• The economy grew by over 7% in 2021.
• The government has limited some civil rights to focus on public order and harmony: for example, most public assemblies require a permit.
• Corporal punishment, including the death penalty, can be used alongside imprisonment.

> **REVISION TIP**
>
> You need to know about how human rights are approached differently by different countries in place context.

Democratic freedom

Some superpowers and emerging powers have transitioned from **authoritarian** systems to more **democratic** government. However, the degree of democratic freedom, the protection of human rights, and the degree of freedom of speech varies.

Authoritarian system	Democratic system
China has a one-party authoritarian political system governed by the Chinese Communist Party. Freedom of association, expression and religion, and the right to peaceful protest are restricted.	India has a democratic government and is the largest democracy in the world – it has over 900 million voters. Individual Indian states hold much of the power (a federal system). Recently, progress has been made on increasing freedom of peaceful protest and other human rights. But concerns over gender equality and violence against women still exist.

> **REVISION TIP**
>
> You need to be able to compare how human rights vary within an authoritarian and a democratic system.

KNOWLEDGE

10 Health, human rights and intervention

10.5 Human rights variations between countries

Political corruption

High levels of corruption are a threat to human rights because corruption may subvert the rule of law.

Levels of **political corruption** around the world vary

The **Corruption Perceptions Index** ranks countries by their perceived level of public sector corruption

Denmark, Finland, New Zealand, Norway, Singapore and Sweden are perceived as the least corrupt governments (2022)

Political corruption

Somalia, Syria and South Sudan are perceived as the most corrupt governments (2022)

Corrupt governments divert money away from ensuring basic needs (such as healthcare, education and housing) are met

Corrupt governments are less likely to uphold their moral and legal obligations to protect human rights, and their judicial system may not respect rights and freedoms

Sub-Saharan Africa

In 2022 in Sub-Saharan Africa:

- conflict and security challenges have hindered attempts to make progress in addressing endemic corruption
- human rights activists are targeted and integrity of voting in elections is questioned
- corruption is damaging democracy, security and development.

Sweden

In contrast, Sweden has:

- laws that are effective at tackling corruption
- public institutions that act with integrity and transparency.

REVISION TIP

You need to understand how political corruption in a country can have a negative impact on the human rights of citizens of that country.

Key terms

Make sure you can write a definition for these key terms

authoritarian   Corruption Perceptions Index   democratic
political corruption

10.6 Human rights variations within countries

Variations of human rights within countries

- There are significant variations in human rights within countries. These are reflected in different levels of social development.

- In some countries (e.g. **post-colonial** states) there are groups, defined by their gender and/or ethnicity, who have fewer rights than the dominant group.

- The differences in human rights are often shown in variations in levels of health and education between the groups.

- In recent years, the citizens of many states have demanded **equality**. Progress is taking place at different rates.

Australia

- The colonisation of Australia (1788–1890) dispossessed the Indigenous peoples of their land.

- Now, there are differences in key indicators of quality of life (health, housing, employment, education) between Indigenous peoples and non-Indigenous peoples in Australia.

- School attendance rates are lower for Indigenous peoples than non-Indigenous peoples, usually due to issues with accessing schools in remote areas. Literacy proficiency levels of Indigenous peoples were around 2–3 years of schooling lower than non-Indigenous peoples.

- Indigenous women report experiencing violence at three times the rate of non-Indigenous women, and face more barriers to getting help.

- Australia adopted the United Nations Declaration on the Rights of Indigenous Peoples (UNDRIP) in 2007 to work to improve the rights of Indigenous peoples in Australia.

- 'Close the Gap' is a campaign to close the health and life expectancy gap between Indigenous peoples and non-Indigenous peoples. The government reports that it is on track to meet the aims.

REVISION TIP

Make sure you know how human rights vary within countries and how this affects the social development of different populations.

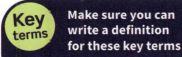

Key terms Make sure you can write a definition for these key terms

equality   post-colonial   United Nations Declaration on the Rights of Indigenous Peoples (UNDRIP)

RETRIEVAL

Learn the answers to the questions below, then cover the answers with a piece of paper and write as many as you can. Check and repeat.

Questions | Answers

#	Question	Answer
1	What is the UDHR?	Universal Declaration of Human Rights
2	Is the UDHR legally binding?	No
3	How were the rights in the ECHR made part of UK law?	By the Human Rights Act of 1998
4	What do the Geneva Conventions protect?	The rights of those engaged in warfare: they are a large part of international humanitarian law
5	How many nations adopted the Geneva Conventions?	196
6	What is the largest democracy in the world?	India
7	What type of government is there in China?	A one-party authoritarian political system governed by the Chinese Communist Party
8	Why has the government in Singapore limited some civil rights?	To focus on public order and harmony
9	How does the UK invoke human rights in international forums?	It is a member of important human rights bodies and presents resolutions and calls to action at international meetings
10	How is the varying level of corruption around the world measured?	The Corruption Perceptions Index ranks countries by their perceived level of government corruption

Put paper here

Previous questions

Now go back and use these questions to check your knowledge of previous topics.

Questions | Answers

#	Question	Answer
1	What does GDP per capita (PPP) measure?	The amount of wealth a country generates within its borders, divided by the population size
2	What is the Happy Planet Index?	An indicator of development that takes into account wellbeing, life expectancy and ecological footprint.
3	Name one IGO that promotes development.	One from: IMF / World Bank / UN / WTO
4	What are the UN's MDGs?	Millennium Development Goals
5	Give one reason why healthcare inequalities exist within developed countries.	One from: lack of access to transport / language barriers / literacy levels / misinformation and poor previous experiences

Put paper here

⚙ KNOWLEDGE

10 Health, human rights and intervention

10.7 Geopolitical intervention

Types of geopolitical intervention

A wide range of **geopolitical interventions** are used to address development and human rights issues.

Development aid in the form of money, technology or expertise is given to one country by another to help improve quality of life

Indirect military action involves military personnel from one country training another country's military, or helping to rebuild or develop infrastructure

Types of geopolitical intervention

Trade embargoes restrict a country from trading with others to put pressure on governments to address certain issues

Direct military action involves military personnel from one country engaging in conflict in another, often forming an alliance with other countries

Military aid in the form of training or weapons can assist countries in conflicts

Interventions by different players

IGOs	The UN promotes and protect human rights through: • peacekeeping missions where a human rights team monitors and investigates issues and supports governments, e.g. UN mission in South Sudan • the Office of the High Commissioner for Human Rights (OHCHR), which supports many countries to protect human rights in line with the UDHR • the UN Security Council, which can vote to impose economic sanctions or support direct intervention. In 2011, the UN authorised a NATO-led military intervention, an arms embargo and a no-fly zone in Libya to protect the human rights of the country's people after the government had failed to do so.
National governments	National governments can intervene directly with another government or appeal for group intervention at international forums, such as the UN Security Council. • Through its government aid agency, the USA provided funds, partnerships and a joint military unit to Honduras. The US claimed their aim was to improve Honduras' justice system, address corruption, and improve standard of living, employment opportunities and democratic governance. • Many governments provide aid with 'strings' attached: they demand that countries who benefit from the aid protect the human rights of their citizens. • The UK government has committed to ensuring that human rights are respected through processes and clauses in the trade agreements it makes with other countries.
NGOs	• Amnesty International is a human rights organisation that campaigns, investigates, and reports on human rights abuses around the world. They have been campaigning to end the use of the death penalty for over 40 years. • Human Rights Watch investigates and reports human rights abuses, partners with other organisations to protect activists, and advises governments, the UN, and other organisations on policy change.

10 Health, human rights and intervention

10.7 Geopolitical intervention

Lack of consensus

There is seldom consensus about the validity of geopolitical interventions.

Views on human rights norms vary around the world.

Intervention in other countries erodes their **national sovereignty** – their right to authority over their own territory.

Lack of consensus on geopolitical interventions

Countries will vote on interventions based on their own geopolitical relationships and interests.

Direct military action can have positive and negative outcomes, although human rights violations are sometimes used as a reason for military intervention.

Libya

The UN justified intervention in Libya in 2011 under its responsibility to protect policy: the UN felt that Libya was not protecting the human rights of its citizens, including their right to freedom of assembly and expression.

However, some countries were concerned that the situation did not justify intervention.

They felt that:

- there was insufficient evidence.
- it would set a precedent for other countries intervening in the way countries decide to treat their citizens.
- intervention is inconsistent: the international community ignores violations in some countries while other countries are subject to embargoes and military intervention.

This action highlighted how intervention to protect human rights was used inconsistently around the world.

> **REVISION TIP**
> You need to know about how governments use condemnation of human rights as a condition for aid and intervention in one place context.

 Key terms **Make sure you can write a definition for these key terms**

geopolitical interventions indirect military action
national sovereignty trade embargoes

10.8 Development aid

Forms of development aid

Development aid can take the form of:
- charitable gifts
- emergency aid to help recover from disasters, e.g. Haiti 2010 earthquake
- specific projects
- loans.

It can be administered by:
- NGOs (e.g. Oxfam)
- national governments
- IGOs.

The impact of development aid

The success of development aid is contested.

Successes	Criticisms
• Progress in dealing with life-threatening conditions. More than 80 countries donate financially to the Global Fund to address HIV, TB and malaria, in line with SDG 3. Since 2022, the fund has saved over 50 million lives.	• Some countries may become dependent on aid from other countries; donations of food aid can undermine the local agricultural economy.
• Improvements in some aspects of human rights. For example, in 2019–20, 45% of OECD development assistance was dedicated to gender equality and women's empowerment.	• **Top-down aid** given to corrupt governments risks being misused for the personal gain of political elites. The World Bank has identified corruption as a major hurdle in eliminating extreme poverty by 2030.
• USAID's 'Feed the Future' programme aims to improve nutrition levels and tackle hunger in 20 countries around the world. It generated $28 billion of agricultural sales between 2011 and 2022 and is estimated to have resulted in 5.2 million families no longer suffering from hunger.	• It is difficult to track how top-down development aid makes a real difference to the standard of living and opportunities of individuals and communities.

Development projects and the environment

Aid and development projects should be sustainable and take environmental considerations into account. Sometimes, programmes led by superpowers and TNCs have very serious impacts on the environment in which minority groups live and disregard their human rights to their land and culture.

> **REVISION TIP**
>
> You need to know about how economic development disregards human rights and the environment in one place context.

Oil in the Niger Delta

- In 2017, oil from the Niger Delta accounted for over 80% of Nigerian government income.
- The area suffers many oil spills due to lack of maintenance, accidents, and purposeful damage.
- Communities in the Delta region suffer from polluted water, loss of farmland, loss of mangrove forests and a decline in biodiversity.

- These ecosystems are important to the way of life of the local population, who have lost basic human rights, such as access to necessities and work.
- Oil spills in 2008–2009 destroyed the fishing industry in Bodo village, Nigeria. TNC Shell offered only $4000 compensation to affected communities before it was taken to court.
- In 2011, Shell reported spilling 17.5 million litres of oil.

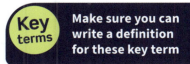
Key terms Make sure you can write a definition for these key term

development aid   top-down aid

KNOWLEDGE

10 Health, human rights and intervention

10.9 Military aid and intervention

Global strategic interests

Global strategic interests might drive **military interventions** but are often justified by the protagonists in terms of human rights.

Libya

NATO-led military intervention in Libya was justified as humanitarian intervention to protect civilians. However, some argue that the international community wanted to intervene for regime change and NATO's bombings were, therefore, unjustified.

REVISION TIP

You need to know about military intervention into human rights issues in one place context.

Military aid

Military aid can include training military personnel and weapons sales.

Military aid is sometimes used to support countries that have questionable human rights records. Countries who send the aid argue that aid will help the receiving countries to protect and promote human rights, and that reducing aid will only exacerbate the situation.

Colombia and the USA

Through 'Plan Colombia' the USA provided financial and military aid to Colombia between 1999 and 2015, despite Colombia's poor record of human rights abuses. The USA argued that it could help Colombia maintain peace and security by attaching conditions to the aid. Amnesty International criticised the aid payments, arguing that not enough aid was withheld if these conditions were not met.

REVISION TIP

You need to know about how military aid is used to support countries with questionable human rights records in one place context.

Military intervention and the 'war on terror'

The 'war on terror' often involves direct military action.

The reason often given for such direct military action is the promotion of human rights, but the actions of combatant states can sometimes compromise this.

Afghanistan

After the 9/11 attacks, US military action in Afghanistan was justified through reports of the Taliban protecting members of al-Qaeda. However, this was compromised by countries that had signed the UDHR using torture. There were reports that terrorists detained by the USA in Guantanamo Bay were subject to human rights abuses such as torture and detention without trial.

REVISION TIP

You need to know how direct military intervention is used in the 'war on terror' but compromised by the use of torture in one place context.

Key terms Make sure you can write a definition for these key term

military aid   military intervention

RETRIEVAL

Learn the answers to the questions below, then cover the answers with a piece of paper and write as many as you can. Check and repeat.

Questions

Answers

Put paper here

#	Question	Answer
1	What are the different types of geopolitical interventions used to address development and human rights issues?	Development aid, trade embargoes, military aid and military action
2	How does the UN promote and protect human rights?	Through the UDHR, OHCHR, the Security Council and peacekeeping missions
3	Name one NGO that reports on human rights abuses around the world.	One from: Amnesty International / Human Rights Watch
4	What are the two main criticisms of development aid?	Aid dependency; corruption and personal gain for political elites
5	Who can aid be administered by?	Governments, NGOs and IGOs
6	What does indirect military action involve?	Military personnel from one country training another country's military or helping to rebuild or develop infrastructure
7	What does direct military action involve?	Military personnel from one country engaging in conflict in another, often forming an alliance with other countries
8	State two forms that development aid can take.	Two from: charitable gifts / emergency aid for disaster recovery / specific projects / loans
9	What is a trade embargo?	Restricting a country from trading with others to put pressure on governments to address certain issues
10	What can military aid include?	Training personnel or supplying weapons

Previous questions

Now go back and use these questions to check your knowledge of previous topics.

Questions

Answers

Put paper here

#	Question	Answer
1	What is the UDHR?	Universal Declaration of Human Rights
2	What do the Geneva Conventions protect?	The rights of those engaged in warfare: they are a large part of international humanitarian law
3	What type of government is there in China?	A one-party authoritarian political system governed by the Chinese Communist Party
4	What is the largest democracy in the world?	India
5	Is the UDHR legally binding?	No

 **KNOWLEDGE**

10 Health, human rights and intervention

10.10 Measuring the success of geopolitical intervention

Measuring the success of geopolitical interventions

Health indicators, such as increasing life expectancy and lowering of infant mortality

Educational indicators, such as an increase in literacy rate and average years of schooling

Increased gender equality, such as lower maternal mortality rate and increased school enrolment of girls

The success of geopolitical interventions is measured by:

The degree that **freedom of speech**, freedom of the press and freedom of information is promoted

The humane treatment and management of refugee flows

The introduction of **democracy**, good governance and political stability

Economic indicators such as an increase in GDP per capita and a decrease in the poverty rate

Democracy as a measure of success

For many countries and NGOs, promoting democracy and the introduction of democratic institutions is an important goal of geopolitical intervention. Democracy can drive other economic improvements and social changes, decreases the likelihood of war, and makes it easier to build economic and military links between countries.

It therefore provides assistance at elections in around 60 countries a year and its peacekeeping missions aim to strengthen democratic institutions. The UN strongly believes in the link between democracy, development and respect for human rights.

Freedom of expression

- Freedom of expression includes the right to hold opinions and communicate to information and ideas.
- It is one of the rights protected in the UDHR.
- It allows criticism of governments, so is important for democracy.

Economic growth as a measure of success

For some countries, success is measured in terms of economic growth. In such cases, less attention is paid to other aspects of **holistic development** (human wellbeing), human rights, or the development of democracy.

Ecuador
Ecuador, which became a democracy in 1979, saw strong economic growth between 2007 and 2017 due to investment in infrastructure and social spending by the government. However, a 2022 report by the US government found human rights abuses, such as restrictions on freedom of expression and the media, and corruption.

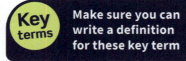 **Key terms** Make sure you can write a definition for these key term

democracy   freedom of speech   holistic development

10.11 Measuring the success of development aid

Successes and failures of development aid

- Aid has a mixed record of success.
- The relationship between aid, development, health and human rights is unclear.

Success in Botswana	Failure in Haiti
<ul><li>Botswana invested income from its natural resources in infrastructure, health, and education.</li><li>Much of Botswana's aid has been focused on fighting HIV/AIDS. Between 2010 and 2019, HIV incidence and AIDS mortality has decreased.</li><li>Botswana has strong governance and low levels of corruption.</li><li>There have been improvements in mean years of schooling and GDP per capita.</li><li>Botswana is expected to no longer qualify for official development assistance from OECD countries by 2030.</li></ul>	<ul><li>Development aid has failed to reach its goals in Haiti.</li><li>Political elites have used money to their advantage instead of developing the country.</li><li>Haiti remains the poorest country in the Western hemisphere, and poverty levels have remained high.</li><li>There is high aid dependency.</li><li>Haiti is still extremely vulnerable to natural hazards and disease outbreaks.</li></ul>

Economic inequalities

- Sometimes **inequality** in states receiving large amounts of aid can increase. In other states receiving substantial aid, inequality can decrease.
- This could be due to corruption, decisions by donor countries on how to allocate aid, or aid agencies focusing on large projects rather than smaller projects that would be effective but would generate less publicity.
- High rates of inequality can lead to lack of access to healthcare and a decrease in life expectancy for some communities.

The **Gini Index** measures income inequality within a country, with a score of zero representing perfect equality and 100 representing maximum inequality.

- Despite rising in the 1990s, Botswana's score fell from 64.7 in 2002 to 40.7 in 2019.
- Haiti's score was 53.3 in 2019.
- The UK's score was 32.6 in 2020.

> **REVISION TIP**
>
> You need to know how development aid can be a success in one place context, and how it can be seen as a failure in another.

Development aid as an extension of foreign policy

Some superpowers use development aid as an extension of foreign policy, to improve their access to resources, geopolitical relationships, and alliances.

- China offers interest-free loans and grants as development aid. It also offers development financing.
- Little information is released about the value and recipients of the aid.
- Much of the aid is focused on sub-Saharan Africa and supports infrastructure projects rather than health, education, and humanitarian aid.
- Some argue that China invests in development to secure natural resources, benefit trade and increase military influence.

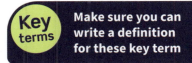 **Key terms** Make sure you can write a definition for these key term

Gini index   inequality

10 Health, human rights and intervention

10.12 Measuring the success of military and non-military intervention

The mixed success of military intervention

Recent direct and indirect military interventions suggest that there are benefits but also significant costs, including loss of sovereignty and human rights.

There are distinct contrasts between any short-term gains and the long-term costs.

Military intervention in Afghanistan

Between 2001 and 2021, US-led military intervention in Afghanistan aimed to remove the Taliban from power, prevent them from harbouring terrorists, and introduce democracy in the country.

Short-term gains of military intervention in Afghanistan

During the intervention, many cities saw more girls going to school and increased female participation in public life. There was a 25% quota of seats for women in parliament.

Short-term gains of military intervention in Afghanistan

There was a fall in the infant mortality rate.

There were improvements in the freedom of the media which allowed increased freedom of expression.

Long-term costs of military intervention in Afghanistan

After the withdrawal of military personnel from Afghanistan in 2021, the Taliban took control of the country.

Long-term costs of military intervention in Afghanistan

Women in Afghanistan now have severely restricted freedoms: girls are banned from secondary school and higher education; women cannot work for the UN, and they must be accompanied by a male relative if travelling more than 72 km.

Estimates suggest that the war cost the USA $2.3 trillion and the lives of 2,448 US soldiers.

4 million Afghans were internally displaced.

There were many civilian deaths from military activities, such as air strikes on Taliban forces.

Success of non-military intervention

Non-military interventions may have a stronger record of improving both human rights and development than military intervention.

Timor-Leste

> Timor-Leste was colonised by the Portuguese. After declaring independence in 1975, it was invaded by Indonesia.

↓

> During Indonesian occupation, many people were killed, displaced or died from hunger and illness.

↓

> The UN supported Timor-Leste's route to democracy independence from Indonesia and through a peacekeeping operation between 1999 and 2002.

↓

> After Indonesia withdrew, the UN maintained law and order, helped develop social services, and assisted in a democratic independence vote.

↓

> The UN is now supporting Timor-Leste in achieving the SDGs.

Impact of lack of action

Lack of intervention may have global consequences and have a negative impact on environmental, political and social development.

- Critics of the UN Security Council argue that it fails to act in situations where action is needed to prevent violence and protect innocent citizens.
- Countries with permanent seats on the Council can use their **veto power** to block any **resolutions** they do not agree with, which may result in a lack of action.

Zimbabwe

- Zimbabwe has a history of human rights abuses, including against those opposing the government of Robert Mugabe (president from 1980–2017).
- Zimbabwe was colonised by the British and regained independence in 1980; some countries may be hesitant around intervention linked to colonialism.
- Some neighbouring countries did not consider Mugabe a threat. Other countries were unwilling to intervene without the support of Zimbabwe's neighbours.
- In 2008, Russia and China voted against a UN resolution for an arms embargo, a travel ban, and financial freeze on the assets of the president and senior government officials.
- Lack of action in Zimbabwe has led to high poverty rates, low life expectancies, high levels of corruption, and high levels of deforestation.

> **REVISION TIP**
>
> You need to know about:
> - the gains and costs of a recent military intervention
> - the success of non-military intervention
> - how lack of action has global consequences in specific place contexts.

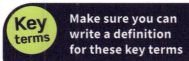

Key terms Make sure you can write a definition for these key terms

resolution   veto power

RETRIEVAL

Learn the answers to the questions below, then cover the answers with a piece of paper and write as many as you can. Check and repeat.

Questions

		Answers
1	Name three ways in which the success of geopolitical interventions can be measured.	Three from: health indicators / education levels / gender equality / freedoms / economic growth / refugee management / more democracy
2	What does the Gini Index measure?	The level of income inequality within a country: a score of 0 represents perfect equality
3	What are the long-term costs of military intervention?	Civilian deaths, displacement, instability, high human and economic cost to combatant states
4	How can development aid be an extension of foreign policy?	It can help improve a country's access to resources, geopolitical relationships, and alliances
5	Why are the permanent members of the UN Security Council powerful?	They can use their veto power to block any resolutions they do not agree with
6	Give one reason why development aid can be seen as a failure in Haiti.	One from: poverty levels remain high / still dependent on aid / still vulnerable to disasters and disease
7	Give two reasons why economic inequality can sometimes increase in states receiving development aid.	Two from: corruption / decisions by donor countries on how to allocate aid / aid agencies focusing on large projects
8	Why might some superpowers use development aid as an extension of their foreign policy?	To improve their own access to resources, geopolitical relationships, and alliances
9	Give two consequences of lack of intervention in Zimbabwe.	Two from: high poverty rates / low life expectancies / high levels of corruption / high levels of deforestation

Put paper here

Previous questions

Now go back and use these questions to check your knowledge of previous topics.

Questions

		Answers
1	What can military aid include?	Training personnel or supplying weapons
2	Who can aid be administered by?	Governments, NGOs and IGOs
3	What are the two main criticisms of development aid?	Aid dependency, corruption and personal gain for political elites
4	What is Amnesty International?	A human rights organisation that campaigns, investigates and reports on human rights abuses
5	What is a trade embargo?	Restricting a country from trading with others to put pressure on governments to address certain issues

Put paper here

PRACTICE

Exam-style questions

1 Explain why education is central to economic development. **(4)**

2 Explain why life expectancy varies in the developed world. **(4)**

3 Explain the role of governments in promoting economic and social development. **(4)**

> **EXAM TIP**
>
> These 4-mark questions have all 4 marks for AO1, so explain your knowledge and understanding of the concepts in the question.

4 Explain how the Universal Declaration of Human Rights (UDHR) protects human rights. **(4)**

5 Explain the long-term costs of military intervention. **(4)**

> **EXAM TIP**
>
> These 6-mark questions have 3 marks for AO1 (knowledge and understanding of the concept(s) in the question) and 3 marks for AO2 (interpretation and analysis of the geographical information in the figure).

6 Explain why democracy is important in terms of development and human rights. **(4)**

7 Study **Figure 1**. Suggest reasons for the variation in life expectancy in English regions 2015-2020. **(6)**

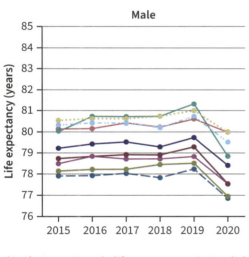
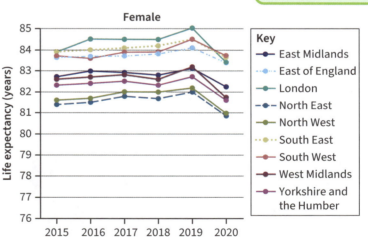

▲ **Figure 1** *Female life expectancy in English regions, 2015 to 2020*

8 Explain how the views of IGOs on development have changed over time. **(8)**

9 Explain why there are significant variations in human rights within countries. **(8)**

10 Explain why there is seldom consensus about the validity of geopolitical interventions. **(8)**

11 Explain how the success of geopolitical interventions can be measured. **(8)**

12 Study **Figure 2**. Suggest why there is a relationship between life expectancy and the percentage of people using safely managed drinking water services. **(8)**

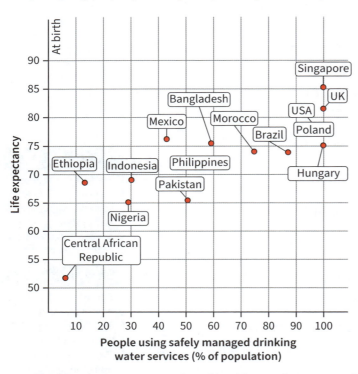

▲ *Figure 2* *Life expectancy and total health spending as a % of GDP (2010)*

13 Study **Figure 3**. Explain why countries vary in both their definitions and protection of human rights. **(8)**

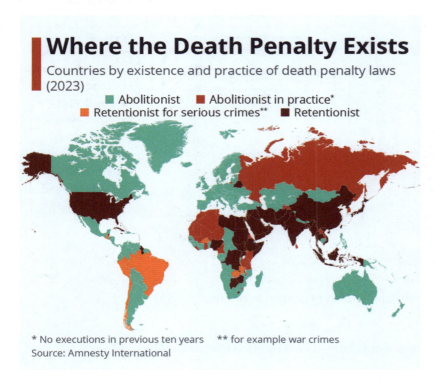

▲ *Figure 3* *Countries by existence and practice of death penalty laws (2023)*

14   Explain why lack of action in development and human rights issues has global consequences.   **(8)**

EXAM TIP

These 8-mark questions have all 8 marks for AO1, so explain your knowledge and understanding of the concepts in the question in detail and use place specific examples where possible.

15   Evaluate the view that the notable variations in human health and life expectancy result from deprivation and income inequalities.   **(20)**

EXAM TIP

These 20-mark questions have 5 marks for AO1 (knowledge and understanding of the processes and concepts in the question) and 15 marks for AO2 (creating a balanced argument, well-evidenced conclusion).

16   Evaluate the view that the United Nation's Millennium Development Goals (MDGs) were unsuccessful in progressing economic and social development.   **(20)**

EXAM TIP

Evaluating a view involves reviewing both sides of the argument and bringing it together to form a conclusion. Points should analyse evidence such as strengths, weaknesses, alternatives, and relevant examples.

17   Evaluate the extent to which development aid is effective in addressing development and human rights issues.   **(20)**

EXAM TIP

Make sure to show both sides of the argument.

Use examples and place contexts that you have studied to exemplify your points.

18   Evaluate the view that development focused on improving both human rights and human welfare has only positive impacts.   **(20)**

11 Migration, identity and sovereignty

11.1 Globalisation and migration

Globalisation and labour demand

Globalisation has caused an increase in migration across many parts of the world. Migration occurs between countries (**international migration**) and within countries (**internal migration**) and can be rural to urban, urban to rural, rural to rural or urban to urban.

Migrants may move for negative reasons in their place of origin (**push factors**) or positive reasons linked to their destination (**pull factors**).

Globalisation has:

- has increased the demand for labour in different regions and within different economic sectors

- increased the number of manufacturing jobs within MICs, particularly in Asia and LICs, leading to rural to urban migration

- resulted in concentrations of high-wage, quaternary sector employment in HICs, especially in urban areas, attracting both internal and international migrants.

Internal migration: China

About a third of China's economically active population (292 million people) are **rural–urban migrants** who now work in China's cities.

Reasons for rural to urban migration in China

- China's open door policy increased international trade and stimulated manufacturing growth in coastal regions, e.g. Shanghai and Guangdong province.

- Lower average incomes, poverty and fewer career opportunities, particularly for young people in rural areas, push internal migrants towards urban areas.

- Decreased restrictions for hukou policies (household registration system). Since the 1990s, local governments have been able to set their own rules to attract workers to fill the **skills gap**. Recent reforms to the *hukou* (household registration) system aim to encourage rural migrants to settle permanently in urban areas.

- Government policy to promote movement from eastern provinces (majority Han Chinese) to north west Xinjiang province (minority Uyghur).

> **REVISION TIP**
>
> Remember, you need to know how globalisation has increased rural to urban and international migration.

International migration: the EU Schengen Area

- The **Schengen** Area is the largest free-movement area in the world.

- The border-free Schengen Area guarantees free movement to more than 425 million EU citizens, along with non-EU nationals living in the EU. Free movement allows every EU citizen to travel, work, and live in any EU country without special formalities.

- This brings economic benefits (including migrants filling job vacancies in other EU countries) and boosts tourism.

- Improves communication and cooperation between national police forces when tackling issues such as illegal migration, serious crime and terrorism.

- Internal border controls can be reintroduced by member states in exceptional circumstances.

Migration policies differ between governments

Governments have different policies for migration depending on the needs of their economy.

Singapore (HIC)	
Population	5.92 million
Key economic challenge	Need for low-skilled, medium-skilled, and skilled labour.
Government policy on immigration	• Two-tiered system for migrant entry linked to employment status: — High skilled – more open policy, encouraged to bring families and have similar rights to residents. — Low skilled – strict social and economic conditions and expected to leave Singapore when job contract expires. • Permission to work managed by a pass system linked to salary/qualifications/work experience: Employment Pass, Work Pass, Skilled Pass, Domestic Worker Pass.
Effects of policy	• Immigration policy a key factor in Singapore's economic growth. • Most Singapore residents are of migrant origin.

REVISION TIP

Write down from memory the main aspects of one government's immigration policy to see how much you can remember.

Patterns of international migration

Differences in migration between regions are caused by different levels of economic opportunity.

Environmental, economic, and political events affect both the source and destination of migrants. These events lead to variations in the flows of:

- **Economic migrants**: People who choose to move from their normal place of residence for economic reasons e.g. work opportunities, favourable tax rules.

- **Refugees**: People who are forced to flee their homes due to war, violence, conflict, or persecution and have crossed an international border to find safety in another country.

- **Asylum seekers**: A person whose request for sanctuary has yet to be processed.

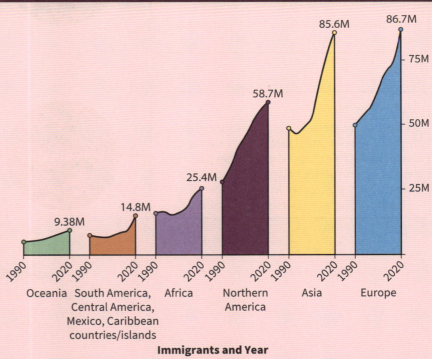

▲ **Figure 1** The number of international migrants has increased globally and across all regions. The greatest increases have taken place in Europe and Asia.

 **KNOWLEDGE**

11 Migration, identity and sovereignty

11.1 Globalisation and migration

Impacts on source and destination areas

The pattern of international migration will continue to change because environmental, economic and political events affect both the **source areas** and **destination areas** of many migrants.

 **Environmental**

Impact on source area:

Global warming can increase desertification and reduce the agricultural productivity through increased variations of rainfall patterns. People migrate to seek a food supply and income.

Impact on destination area:

Increase in low skilled migrants, some of whom may have arrived illegally. Destination countries may have to cope with large numbers of climate refugees.

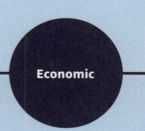 **Economic**

Impact on source area:

Job losses can lead workers to seek jobs abroad. They may send remittances home. Migration of professionals may result in a 'brain drain'. Economic disparities within countries lead to some workers moving to regions with more stability and prospects.

When there are fewer opportunities abroad, workers are more likely decide to stay.

Impact on destination area:

Areas less affected by crisis are more attractive to migrants. Immigration rules may be relaxed to facilitate migration and/or make it more selective.

 **Political**

Impact on source area:

Short-term increase in refugees. Outmigration might be gender imbalanced (more males than females) if country conscription policies demand men. Internal displacement of people within country increases in times of conflict (3.7 million displaced persons in Ukraine).

Impact on destination area:

Inter-regional and international forced migration. Bordering countries may be first option for refugees, putting pressure on housing and services in these areas. More wealthy /liberal countries may open up borders to accommodate refugees more quickly. There are 6.2 million refugees from Ukraine globally.

> **REVISION TIP**
>
> Draw a table to show the impacts of migration on both the source and destination countries to help you remember the key facts.

 Key terms

Make sure you can write a definition for these key terms

asylum seeker   destination area   economic migrant   globalisation
internal migration   international migration   push factors
pull factors   refugee   rural-urban migration
Schengen Agreement   skills gap   source area

11.2 Causes of migration

Common causes of migration

Most migrants are economic migrants who move for work or to rejoin family members who have already moved to another location.

Displaced persons are those who move outside their usual place of residence, often due to conflict, violence, poverty or disasters. This includes refugees. Over 59 million people are currently displaced within their own countries.

Migrants crossing the Mediterranean

- Every year, thousands of migrants, including refugees, cross the Mediterranean Sea to reach Europe.
- Migrants come from countries affected by war and conflict, e.g. Syria and Afghanistan, or are economic migrants who hope to find employment and a better quality of life.
- It is a dangerous crossing. Migrants may pay smugglers a fee to travel in overcrowded and unseaworthy boats, with limited safety equipment.
- Spain, Italy and Greece, which are EU member states, have received the most migrants crossing the Mediterranean and have received additional assistance from the EU to help support arrivals.
- NGOs, including humanitarian groups, help with search and rescue and provide support to arrivals.
- Migrants may face problems on arrival in Europe including discrimination and lack of suitable accommodation.

> **REVISION TIP**
>
> Write flashcards with as many reasons for migration (push and pull) as you can. Avoid 'mirror answers' where you repeat the same point by phrasing it in a positive and negative way: e.g. PUSH fewer employment opportunities; PULL more employment opportunities.

Freedom of movement

Neo-liberalism is an economic theory that suggests that **economic efficiency** is increased when there is free movement of goods (free trade), capital (deregulated financial markets) and labour (**open borders**).

Within the EU's single market, the 'four freedoms' govern the movement of goods, people, services, and capital, with the aim of bringing increased economic benefits to member states.

> **REVISION TIP**
>
> For each of the following, record an audio file listing the impacts of their free movement on national identity and sovereignty: goods, capital and labour.

Benefits of free movement	Costs of free movement
<ul><li>Allows the economy to react to supply and demand.</li><li>Skilled workers can be obtained more easily, filling labour shortages and developing businesses.</li><li>Allows for investment and innovation in businesses and regions where it is most productive.</li><li>Increases cultural exchange, diversity and knowledge transfer.</li><li>A larger pool of workers helps issues linked to ageing populations.</li><li>Increased competition leads to reduced prices and more consumer choice.</li><li>Some see freedom of movement as a fundamental human right and the opportunity to be a 'global citizen'.</li><li>Economic benefits of free movement to 'core' regions with 'trickle-down' to peripheral regions.</li></ul>	<ul><li>Erodes **national sovereignty** as countries have less control over their own borders.</li><li>Challenges **national identity** – without borders, what is a **nation state**?</li><li>Increased labour supply can reduce wages, especially for low-skilled and local workers.</li><li>Culture clashes can lead to tensions within societies.</li><li>Limitless migration can put pressure on key social services, e.g. health and education.</li><li>Increased regional disparities as workers more likely to move to 'core' than peripheral regions.</li><li>Concerns over border controls and security.</li><li>Rise of nationalist and populist parties.</li></ul>

11 Migration, identity and sovereignty

11.2 Causes of migration

Internal migration

Internal migration in many states is unrestricted. This helps to ensure efficient allocation of resources.

Internal migration within the UK

Movement within the UK, including for work, is unrestricted by the government.

In the UK, many of those who move are young (between 20 and 35) and move to urban areas. Many older internal migrants move to rural or coastal areas.

Regional migration in the UK

In 2015:

- Regions in the south of England had a greater net change than regions in the north of England.
- London experienced the highest negative net internal migration levels.
- Eastern and Southern England had a positive net internal migration flow (highest in the South West).
- The North East had the smallest inflows and outflows.

Since 2020:

- London has experienced the highest net outward migration levels.
- Urban areas outside of London have also experienced net outward migration.
- Urban areas with large nearby rural populations have experienced stable net inward migration levels.
- Predominantly rural areas have experienced higher net inward migration levels.

Global migration

Some global regions have unrestricted migration:

- The Schengen Agreement enables EU citizens to move freely within EFTA and most EU countries.
- The Central America-4 Free Mobility Agreement allows free movement between Guatemala, Honduras, El Salvador and Nicaragua.

There is no system allowing free movement between all sovereign nations. Some people perceive themselves as '**global citizens**'. They see themselves as part of a world community whose actions help to define the global community's values and practices. However, this idea is restricted in practice by the rules, regulations, and controls of individual nation states.

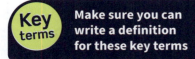

Key terms — Make sure you can write a definition for these key terms

displaced person   economic efficiency   global citizen
national identity   nation state   national sovereignty   open borders

11.3 Consequences of migration

Cultural and ethnic composition of countries

Migration can change the **cultural and ethnic composition** of countries.

Migration can enrich cultural and ethnic diversity. Some migrants may create or gravitate to areas with populations who have similar backgrounds.

Some migrant groups may find **integration** more difficult. Cultural conflicts can arise due to prejudice, misunderstanding or competition for services and resources.

> **REVISION TIP**
>
> Write out a list from memory of how migration influences the cultural and ethnic make-up of countries.

Political tensions

Migration can affect voting patterns as different groups have different priorities.

UK	USA
In the 2016 UK EU membership referendum, some people voted to leave the EU (Brexit) because of the EU's policy on freedom of movement of people and the perceived loss of decision-making powers to EU institutions.	Since 2000, the size of the immigrant electorate in the USA has nearly doubled to 23.2 million. Most immigrant eligible voters are Latinx or Asian.

Perceptions of the impact of immigration

Individuals and groups have different **perceptions** of the impacts of migration. These views may change over time. Perceptions of migration can differ between migrants themselves, people in the destination country, and the population of the source country.

	Positive perceptions of migration	Negative perceptions of migration
Social	• Promoting understanding and tolerance between different groups • Contributes to community life	• Culture clashes • Tensions between groups • Challenges of integration
Economic	• Fills labour shortages and skills gaps • Sending money to source country • Entrepreneurial spirit	• Increased competition for jobs • Low-skilled migrant workers may lead to lower wages in certain sectors • Cost of services and welfare (e.g. health, education)
Cultural	• Increased cultural diversity • Introduces new ideas e.g. food • Enriches arts and entertainment	• Loss of culture in the destination country • Religious and cultural differences may lead to misunderstandings and conflicts • Perceptions surrounding security and crime
Demographic impact	• Can offset declining birth rates • Youthful workforce	• Pressure on housing and infrastructure • Outmigration, loss of young people and 'brain drain' in source country

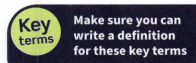

Key terms Make sure you can write a definition for these key terms

cultural and ethnic composition   demographic impact
integration   perception

⚙ KNOWLEDGE

11 Migration, identity and sovereignty

11.3 Consequences of migration

Labour flows across the Mexico–US border

Migrants from Mexico include high- and low-skilled workers, as well as illegal migrants.

Positive perceptions	Negative perceptions
• Legal (high- and low-skilled) and illegal migrants fill labour shortages e.g. in agriculture, construction and hospitality • Immigration can balance the costs of an ageing population, bringing in youthful workers who pay taxes • Increased multiculturalism, including in the arts and entertainment • In 2022, remittances from the USA to Mexico totalled US$63.2 billion	• Areas with high unemployment may want to prioritise jobs for American citizens • 9/11 increased concerns about national security, leading to a desire for more selective immigration • Perceptions of crime, including illegal drugs and gang violence • Levels of unemployment and poverty in Mexico increase economic migration to the USA • Perceptions of a future 'white minority' in the USA • Pressure on services, including security in border areas

Labour flows between EU states after 2004

- After EU enlargement in 2004, there were large labour flows from eastern to western Europe. Those migrating had all levels of educational background.
- Labour flows from southern Europe to northern Europe were characterised by highly educated people, especially after the financial crisis in 2008.

- Labour flows are not only linked to wage difference but also career opportunities.
- Migrants are likely to be temporary with most wanting to move back to their home country within five years.

REVISION TIP

Make a one-minute audio recording to explain examples of control between international borders, globally and within regions.

Positive perceptions	Negative perceptions
• Working abroad increases the possibility of finding a job or a better, more suitable one • Just under 20% of EU citizens have lived in another EU country or can see themselves working in another country in the future • Higher education levels make people more likely to move as they are better equipped to deal with obstacles, e.g. language differences	• Loss of human capital, especially young talent • Home EU state finances students' university education but other EU countries benefit from students' skills when they move abroad

Obstacles to migration

- Income levels – poorer migrants cannot afford the cost of living at the destination.
- Literacy levels – less literate migrants may find it harder to find employment.
- Skills levels – demand for skilled jobs varies in different sectors of the economy.
- Employment availability – job shortages may lead to preferential entry for some workers.
- National policy – countries may impose stricter controls on some groups of immigrants than others.

- Conflict – migrants may not be allowed to move to conflict zones or may find it hard to move out of them.
- Bureaucracy – migrants may find it hard to navigate immigration policies, e.g. passports and visas.
- Misinformation – perceptions of a place may be different to reality.
- Moving costs – migrants may find it difficult to afford transport and removal costs as well as set-up costs at destination.

RETRIEVAL

Learn the answers to the questions below, then cover the answers with a piece of paper and write as many as you can. Check and repeat.

Questions

		Answers
1	What is international migration?	Migration across an international border
2	What is internal migration?	Migration within a country
3	What is a push factor?	A negative reason for moving away from a place of origin (source area)
4	What is a pull factor?	A positive reason for moving to a new location (destination area)
5	Give one cause of rural to urban migration in China.	One from: lower average incomes / poverty and fewer career opportunities in rural areas / open door policy created manufacturing jobs in urban areas / decreased restrictions for hukou policies / government policies promoting movement from eastern provinces to northwest province
6	What is the Schengen Area?	A free travel area covering 27 European countries
7	What is a displaced person?	Someone who moves outside their usual place of residence due to negative factors such as conflict or disasters
8	Name the four freedoms within the EU.	Goods, people, services and capital
9	Name one regional-level free movement agreement.	One from: Schengen Agreement / the Central America-4 Free Mobility Agreement
10	Give two obstacles to migration.	Two from: income levels / literacy levels / skills levels/ employment availability / national policy / conflict / bureaucracy / misinformation / moving costs
11	What is the skills gap?	Shortages in particular employment areas that often cannot be filled by the existing economically active population
12	What is a labour flow?	Movement of economic migrants from one place to another for employment

Put paper here

 **KNOWLEDGE**

11 Migration, identity and sovereignty

11.4 Nation states

National sovereign states

Nation states are **sovereign territories** that are recognised as independent states by their citizens, other countries, and the United Nations. Sovereignty is the authority of a place and its people to govern itself.

There are great differences in the ethnic, cultural and linguistic composition of different nation states and within their borders. The **homogeneity** of a country depends on the history of their population growth, their interaction with other nation states, and the influence of migration.

> **REVISION TIP**
>
> Remember, you need to know how and why the ethnicity and culture of national sovereign states varies.

Iceland: a homogeneous population

- Iceland has a population of approximately 387,000.
- Until the mid-1990s, over 90% of Icelandic citizens were ethnic Icelanders of Norse or Celtic ancestry. Those gaining Icelandic citizenship had to adopt an Icelandic name.
- Approximately 93% of Icelanders speak Icelandic, the official language, as their first language.
- Icelanders are mainly Christian (approximately 62% are Lutherans).

Factors causing homogeneity
• Island nation in the North Atlantic, so geographically isolated.
• Limited immigration.
• However, following the banking crisis in the late 2000s, Iceland's population has become more diverse due to increased immigration.

Singapore: a diverse population

- Singapore has a population of approximately 5.95 million.
- Singaporean citizens mainly come from three ethnic groups: Chinese approximately 74%, Malay approximately 14%, and Indian approximately 9%.
- Within each main ethnic group, there are different languages, customs, and religions (including Buddhism, Christianity and Islam).
- Singapore has four official languages (English, Malay, Mandarin Chinese and Tamil) reflecting the country's history of migration.

Factors leading to diversity
• Geostrategic trading location in SE Asia.
• History of migration with overseas labour an important part of the economy.
• Colonised by the British and regained independence in 1965.
• Ethnic composition of Singapore's citizen population.

National borders

National borders have developed over time. Many national borders are a consequence of physical geography, historical development, and colonial history, and some are disputed. Borders may not take account of different ethnic, religious or linguistic groups. This can cause problems of sovereignty, recognition and **legitimacy**.

Iraq border

Modern political instability and ongoing conflicts in the region have been linked to the Sykes-Picot Agreement, drawn up by the UK and France to divide the Middle East under British and French protection and rule in 1916.

- Some groups became minorities in the new countries, leading to tensions.
- Sunni and Shia Muslim communities were split.
- The Kurd ethnic group was divided into four regions. Some Kurdish nationalists want to create a nation state, while others seek more autonomy for their population within existing countries.
- Daesh, a terrorist organisation, sought to create its own state following political instability caused by the civil war in Syria and the withdrawal of allied forces.

> **REVISION TIP**
> Draw a memory map to show the physical and human ways different national borders have been designated.

Contested borders

Ukraine and Russia

The border between Russia and Ukraine has changed over time. Crimea is a strategic location as it provides a warm-water port that gives naval and commercial shipping access to the Mediterranean via the Black Sea.

> **REVISION TIP**
> Annotate a map of a country that has contested borders with facts to explain why the borders are contested.

Crimea was incorporated into Russia.

Ukraine gained independence following the breakup of the Soviet Union.

Russia invaded Ukraine. Millions of people left Ukraine. The conflict has led to tensions between Russia and NATO, and economic sanctions have been placed on Russia.

1783 > **1954** > **1991** > **2014** > **2022**

Crimea was transferred to the Ukraine Soviet Socialist Republic (part of the USSR).

Russia annexed Crimea. Conflict took place between Russia and Ukraine in the Donbas region (Eastern Ukraine region with a 38% ethnic Russian minority and significant coal reserves).

Taiwan

- Since 1949, China has claimed sovereignty over Taiwan, considering it part of China.
- The Chinese government will not have diplomatic relations with countries which recognise Taiwan as an independent country.
- Taiwan adopted an export-orientated economy and has experienced rapid economic growth. It is an advanced technology-based economy.
- China and Taiwan are linked economically. Taiwanese firms invest in China, and China is Taiwan's largest trading partner (25% of total exports).

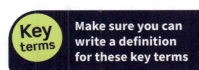 **Key terms** **Make sure you can write a definition for these key terms**

homogeneity   isolation   legitimacy   sovereign territory

⚙ KNOWLEDGE

11 Migration, identity and sovereignty

11.5 Nonalism

Nationalism in the 19th century

Around 1500–1900 CE, European powers colonised numerous countries to build global **empires**. Colonial powers brought their infrastructure, culture and language to lands they invaded. European powers believed they could strengthen their own nation states by colonising others.

> **REVISION TIP**
> Remember, you need to know how 19th century nationalism helped to develop empires.

India under colonialism: The British Raj

> Britain indirectly controlled parts of India through trade arrangements between Indian rulers and the British East India company. Following a rebellion by Indian troops in 1857, the British government brought India under its direct control (The British Raj).

> During the British Raj, there were movements for Indian independence (e.g. Indian National Congress and Mahatma Gandhi).

> The Partition of India, after the dissolution of the British Raj, created the nation states of India, Pakistan and East Pakistan (now Bangladesh).

1857 ＞ **1858** ＞ **Before 1947** ＞ **1947** ＞ **After 1947**

> In 1858, the British government passed the Government of India Act, transferring power to the British crown.

> Indian **nationalism** grew until India gained independence in 1947.

New nation states

Since 1945, many new nation states have emerged as European empires have disintegrated, often causing conflicts. European empires disintegrated more rapidly than they were built. Since the creation of the United Nations in 1945, 80 countries that had been colonised have gained their independence.

After the Second World War, European powers faced issues in several domains:

- Economic: the need to spend money on reconstruction at home rather than on maintaining countries they had colonised.

- Political: the rise of nationalism in colonised countries, and changing attitudes of people within colonial powers, led to opposition to colonialism at an influential level.

- Social: an increasing resistance to foreign rule and the demand for independence within colonies.

> **REVISION TIP**
> List the reasons why many countries became independent in the post-WWII era.

The 'Wind of Change' in Africa (1960s)

- After the Second World War, African countries began to gain independence.

- The British government was concerned that violent conflict would spread to British colonised countries in Africa. Prime Minister Harold Macmillan gave his famous **'wind of change'** speech in 1960. The gradual decolonisation of Africa increased and the 1960s saw many new states emerge.

- Independence movements gathered pace. Countries developed an understanding of how independence

could be achieved by looking at the experience of others.

- In some states, stable governments were not put in place, sometimes resulting in civil war (e.g. in Angola).

- Economic development was affected by the difficult transition to independence.

- Minority groups were sometimes ignored by new governments, leading to resentment and conflict.

The cost of disintegrating empires

Decolonisation has caused conflicts that are costly environmentally, economically, and in human terms.

Sudan (Northeast Africa)

REVISION TIP

Draw a mind map of the impacts of decolonisation on both the former imperial core country and its former colonies.

Under British and British Egyptian rule. Governed as two Sudans: North (predominantly Muslim and Arabic speaking) and South (more diverse).

Political instability, military coups and civil war.

South Sudan becomes independent.

| 1899–1956 | 1956 | 1956–72 | 1980s–90s | 2011 | 2013–20 |

Sudan becomes independent.

Further conflicts between government and Sudan People's Liberation Army.

Civil war. 2020 Juba peace agreement, but internal conflicts continue.

Environmental costs

- Land degradation through unsustainable agricultural practices and military activities, made worse by desertification caused by climate change.
- Water scarcity: Competition for access to water in arid regions causes conflict.
- Loss of biodiversity: wildlife and habitats exploited by military groups to help fund their campaigns.

Economic costs

- Economic development and trade disrupted by damaged infrastructure (including oil facilities).
- Decline in agricultural productivity leading to low crop yields.
- South Sudan's economy is heavily dependent on oil but issues surround sharing revenues with and transport through Sudan.
- High military expenditure leaves less for investment on education and healthcare.

Human costs

- Over 400,000 civilians and military personnel killed.
- Over 4 million people displaced.
- South Sudan has a low HDI.
- Sudan and South Sudan are dependent on development aid.
- Millions face food insecurity in South Sudan.

Patterns of migration from former colonies

- Some former colonies retain links with their former colonisers. Many people have migrated from former colonised countries to the UK, particularly after independence. The British Nationality Act 1948 granted 'subjects of the British Empire' the right to live and work in the UK. Between 1953 and 1962 around half a million people arrived in the UK from outside of Europe, including from the Caribbean, India and Pakistan.

- By the mid-1970s a third of the UK's total number of immigrants came from the 'New Commonwealth'. Economic migrants filled gaps in sectors of the economy with labour shortages, e.g. working as bus drivers, in the textile industry, or in the NHS. The majority settled in London and other large urban areas.
- This movement of people, and subsequent migrations of refugees and economic migrants from other parts of the world, has changed the ethnic composition of the UK, making the country more heterogenous over time.

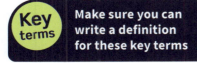

Key terms Make sure you can write a definition for these key terms

empire   nationalism   'wind of change'

 **KNOWLEDGE**

11 Migration, identity and sovereignty

11.6 Globalisation and global inequality

Tax havens

- The globalisation of business activity has led to financial deregulation.
- Businesses have more freedom to operate outside of their own country and take advantage of more favourable financial rules in tax havens.
- This has encouraged the growth of states that have low-tax regimes.
- **Tax havens** are attractive to wealthy individuals who want more protection for their assets, companies wanting to cut costs and maximise profits, and companies and individuals wanting more confidentiality for their investments.

 REVISION TIP
Remember to think about why countries with low-tax regimes have grown.

Arguments for tax havens	Arguments against tax havens
• Accepted by governments and IGOs as part of the global economic system.	• NGOs criticise tax havens for encouraging tax avoidance and therefore limiting flows of capital to fund development, widening global inequality.
• Tax laws enable clients to reduce the taxes they pay or to eliminate their tax liability altogether (e.g. Cayman Islands has zero corporate income tax and no capital gains, personal, or inheritance tax).	• Governments in the 'home' countries of individuals and businesses who use tax havens lose tax revenue, so they have less money to spend on public services.
• Overseas citizens can register 'offshore' companies that have preferential taxes and may not have to pay local taxes.	• Privacy and secrecy can encourage illegal businesses and activities (e.g. money laundering and the illegal drugs market).
• Increased privacy and confidentiality as details do not have to be made public.	• Negative reputation surrounding tax avoidance; home nation's tax authorities may scrutinise companies more carefully.
• Tax havens are legal.	• Perceived risks may put off investors.
• Tax havens can set up accounts quickly making it easy and convenient for the investor.	• Some IGOs have tried to put rules in place to ensure companies pay tax wherever they are based. In 2021, the 'Global Minimum Tax Agreement' was put forward to ensure large companies paid a minimum of 15% tax.
• Assets can be protected as they are outside of major jurisdictions (USA, EU) and do not often have Tax Information Exchange Agreements (TIEAs), making scrutiny more difficult.	

 REVISION TIP
Draw two columns, 'for' and 'against' tax havens, and write down as many points as you can think of for each.

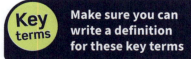 **Make sure you can write a definition for these key terms**

alternative economic model   tax haven

The threat of growing global inequalities

- Global inequality is growing.
- MENA (Middle East North Africa) has the most income disparity. Europe has the lowest income disparity.

- Inequality between the world's highest-income countries and the lowest is rising.
- Inequality within countries is rising, e.g. between urban core and rural periphery.

Global income

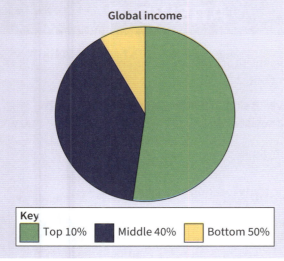

Key
- Top 10%
- Middle 40%
- Bottom 50%

Global wealth

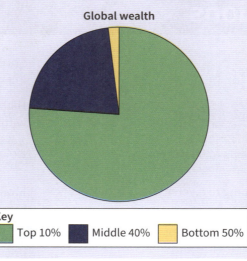

Key
- Top 10%
- Middle 40%
- Bottom 50%

◀ **Figure 1** *The global bottom 50% has 8.5% of total income and 2% of global wealth. The global top 10% has 52% of total income and 76% of global wealth.*

Alternative economic models

Growing inequalities have been identified as a significant threat to the sustainability of the global economic model. Therefore, some governments have promoted **alternative economic models** to the globalised capitalist free-trade model in order to reduce inequality.

Bolivia

- Nationalised oil, gas, and mining, so increasing money earned for the country
- Promotion of other state owned companies
- Increase in public investment
- Money gained used to fund social projects

Socialism of the 21st century

- Welfare programmes to reduce poverty
- Land reforms to increase food security
- GINI coefficient fell from 0.56 (2005) to 0.4 (2018)
- Literacy rate and GDP increased

Bolivia still has challenges:

- problems of bureaucracy and mismanagement
- state firms uncompetitive in the global market, some going bankrupt
- lack of diversification of the economy
- increasing overseas debt (quadrupled between 2006 and 2019)
- lack of overseas investment
- 77% of Bolivians work in the informal sector so may pay less tax to the government.

> **REVISION TIP**
>
> Choose one alternative economic model and practice explaining it to a friend.

Learn the answers to the questions below, then cover the answers with a piece of paper and write as many as you can. Check and repeat.

Questions | Answers

	Questions	Answers
1	What is a nation state?	Sovereign territory recognised as an independent state
2	What is homogeneity?	High level of similar culture, language and beliefs within a society
3	Give two factors leading to homogeneity.	Isolation, limited migration
4	Give two factors leading to diversity.	Strategic trade location, immigration
5	What was the Sykes-Picot Agreement?	1916 agreement dividing UK and French spheres of influence in the Middle East
6	Who gave the 'wind of change' speech in 1960?	British Prime Minister Harold Macmillan
7	What was the British Raj?	The direct militarised control of India by the British government as a result of colonisation
8	When did India gain independence?	1947
9	What is a tax haven?	Country or territory that offers very low or zero tax rates
10	What is the Global Minimum Tax Agreement?	Initiative to ensure companies pay 15% tax regardless of where they operate
11	Which country had the 'Socialism of the 21st Century' policy?	Bolivia

Put paper here

Previous questions

Now go back and use these questions to check your knowledge of previous topics.

Questions | Answers

	Questions	Answers
1	What is international migration?	Migration across an international border
2	What is internal migration?	Migration within a country
3	What is a push factor?	A negative reason for moving away from a place of origin (source area)
4	What is a pull factor?	A positive reason for moving to a new location (destination area)
5	Give one cause of rural to urban migration in China.	One from: lower average incomes / poverty and fewer career opportunities in rural areas / open door policy created manufacturing jobs in urban areas / decreased restrictions for hukou policies / government policies promoting movement from eastern provinces to northwest province

Put paper here

⚙ KNOWLEDGE

11 Migration, identity and sovereignty

11.7 The United Nations

The UN: the first post-war IGO

The United Nations (UN) is an IGO set up in 1945 with 51 original members. It has grown in influence and, in 2023, there were 193 UN member states.

International peace and security

Human rights

Humanitarian aid

Role of the UN

Sustainable development and climate action

International law

UN member states sign the UN Charter which states the purpose of the UN. This is to:

- maintain international peace and security by preventing and removing threats to peace and settling international disputes peacefully
- develop friendly relations between countries based on equal rights and self-determination
- solve international economic, social, cultural or humanitarian problems, promoting **human rights** and freedoms for everyone

- coordinate nations as they work towards achieving these goals.

> **REVISION TIP**
>
> Try to write about as many roles and functions of the UN as you can in a minute. Check your answers and try again.

UN agencies

Issue	Agencies
Environmental	UN Framework on Climate Change UN Department of Economic and Social Affairs (17 sustainable goals)
Socio-economic	UN Development Programme UN Educational, Scientific and Cultural Organization
Political	UN General Assembly UN Security Council

UN Security Council

- The UN Security Council examines threats to world peace and security. It votes on decisions about how to respond to these threats.

- Rotation ensures all countries have a seat on the Security Council.

- Nations on the Security Council have different types of governments, geopolitical views, and priorities. This makes it difficult to reach an agreement on many issues.

- Five permanent members (China, France, Russia, UK, and USA) have the right to veto decisions made by the Security Council.

▲ *Figure 1 The UN headquarters in New York, USA*

⚙ KNOWLEDGE

11 Migration, identity and sovereignty

11.7 The United Nations

Interventions by the UN

The UN can intervene in the defence of human rights in two main ways:

- **Economic sanctions** on a country which is threatening another through restricting trade (embargoes), travel bans, financial and commodity restrictions or interrupting transport and communications.

- **Direct military intervention** is deployment of UN peacekeeping forces (made up of civilians, military and police) to monitor ceasefires, reduce land-mine risk, disarm former military personnel, and protect the law and human rights.

Iran: embargo (2006–2020)

- Embargo on the export of nuclear weapon technology to Iran.
- Embargo on Iran's arms exports and imports, including missiles, combat aircraft, and tanks.
- Aim was to reduce Iran's ability to develop nuclear weapons.
- Security Council voted against USA proposal of extending sanctions.
- Sanctions also affect Iranian civilians.

Democratic Republic of Congo: military intervention (1999 onwards)

- Civil war between ethnic and political rivals.
- Sides supported by neighbouring countries.
- Mission to support democratic elections. Multinational peacekeeping force (MONUSCO) of military and police.
- In 2023, intervention involved 17,700 personnel at a cost of $1 billion per year.
- Some success: elections have been held but there is still combat in eastern DRC.

Unilateral action

- Some UN member states have operated independently to intervene in countries they claim are '**fragile states**', or to conduct a '**war on terror**', disregarding the decisions of the Security Council.

- This threatens relationships between countries, threatens global stability, and weakens the position of the Security Council.

USA and UK military invasions, e.g. Afghanistan (2001), Iraq (2003)

USA drone strikes, e.g. Pakistan (2004)

Non-Security Council approved interventions

USA military intervention in countries labelled as 'fragile states', e.g. Yemen (2022)

REVISION TIP

Remember to think about *why* some countries have acted independently of the UN.

Key terms

Make sure you can write a definition for these key terms

direct military intervention   economic sanctions   embargo   'fragile state'   human rights   United Nations (UN)   UN Security Council   war on terror

11.8 Financial IGOs

International financial organisations

Three IGOs that have been established since the Second World War control the rules of world trade and financial flows.

REVISION TIP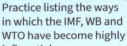

For each of the IMF, WB, and WTO, bullet point what they are and what they do.

Financial IGOs

The International Monetary Fund (IMF, established 1944)	World Bank (WB, established 1944)	World Trade Organisation (WTO, established 1995)
• Encourages global financial cooperation • Promotes increased trade and economic growth • Encourages governments to have policies which increase prosperity • Offers technical assistance and training to governments • Gives loans to countries to help improve their economies	• Gives financial and technical help to countries to increase their level of development, including low interest loans and grants	• Promotes free trade and settles trade disputes between different countries • Puts rules in place to govern how goods, services, and intellectual property are traded around the world

Importance of the IMF, World Bank, and WTO

The IMF, WB, and WTO were established to increase international trade and have been important in maintaining the dominance of 'western' capitalism. The USA has the highest percentage share of votes in the World Bank and the IMF.

Financial help these IGOs give to developing countries may be conditional on the countries putting 'western' capitalist ideas into place, including:

REVISION TIP

Practice listing the ways in which the IMF, WB and WTO have become highly influential.

- Free trade: barriers to trade between countries removed or weakened.
- Market economy: trade of goods and services relies on supply and demand.
- Competition: business and individuals compete with each other in the market to improve efficiency, innovation, and consumer choice.
- Privatisation: private, non-state, ownership of property, businesses and intellectual property.
- Profit-driven initiatives.
- Limited government intervention.
- Entrepreneurship encouraged.
- Democracy: people are free to elect their chosen government.

Global borrowing and trade

Global borrowing rules and trade policies have led to economic growth in the developed world.

Developed countries increase their economic growth by:
- exporting their higher value goods and services to developing countries
- accessing capital to invest in infrastructure and innovation
- benefiting from free trade policies to open up new markets
- benefiting from economies of scale
- FDI (encouraging investment in their own countries and investing in middle- and lower-income countries)

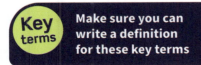
Key terms Make sure you can write a definition for these key terms

EU   global borrowing rules   HIPC   IMF
NAFTA   SAP   World Bank   WTO

11 Migration, identity and sovereignty

11.8 Financial IGOs

Structural Adjustment Programmes (SAPs)

- The IMF and World Bank lent money to increase economic development, but with conditions.
- One condition was that countries had to export more and reduce government spending.
- Cuts to government spending impacted health and education, leading to a slowdown in human development.
- Some argue that the conditions also made countries forfeit their economic sovereignty.

Heavily Indebted Poor Countries initiative (HIPC)

- Introduced by the IMF and World Bank in 1996.
- Countries gain debt relief on loans if they meet set criteria, including poverty-reduction strategies.
- Countries have to show good financial management and lack of corruption.
- Savings must be spent on poverty reduction, health and education.

Jamaica's SAP

Jamaica accepted loans in the 1970s and 1980s.

Conditions of the loans	Impacts
• Decrease in government spending. • Currency devaluation and reducing inflation. • Free trade policies and privatisation. • Low-wage economy to encourage FDI.	• Health: decrease in nurses and increase in mothers dying in childbirth. • Education: decrease in the number of children attending primary school. • Jamaica still has a high debt burden but does not qualify for debt relief through the HIPC initiative.

Global and regional trade

North American Free Trade Agreement (NAFTA 1994–2020)	European Union (EU): 27 member states
• Trilateral trade agreement: USA, Canada, Mexico. • Aims to remove trade and investment barriers and increase investment opportunities. • USA is the largest trading partner of the three nations. • 2020: replaced by the USMCA to include intellectual property and digital services and improve environmental regulations. • Advantage: growth of trade and investment. • Disadvantage: USA exports cheap agricultural products to Mexico, increasing competition for Mexican farmers.	• Economic integration through a single market removes trade barriers and allows free movement of people. 20 EU member states share a currency (euro). • The EU has trade agreements with other blocs and individual countries. • Political integration through the European Parliament and European Commission. Court of Justice of the European Union oversees the application of EU laws.

11.9 Environmental IGOs

Global environmental issues

- The **atmosphere**, **biosphere** and **hydrosphere** are part of the global commons.
- Individual countries have access to resources within the global commons, but long-term sustainability of those resources is improved with the cooperation of all nations.
- Environmental problems linked to the atmosphere and biosphere cross international borders. This makes it difficult for individual nations to solve these issues and to reduce their impact.
- Environmental IGOs have been set up to tackle different global environmental problems, e.g. atmospheric pollution, climate change, reduction in biodiversity, management of water sources and spaces, protecting wilderness.

The atmosphere and biosphere

Montreal Protocol on Substances that Deplete the Ozone Layer (1987)

Limited and eventually banned production and use of CFCs (chlorofluorocarbons). CFCs, used in refrigeration and aerosol sprays, were found to deplete the ozone layer that protects humans and other living things from the Sun's ultraviolet radiation. Depletion of the ozone layer has negative effects on health and the environment.

Successes

- Successful agreement as it was universally ratified.
- The ozone layer has started to recover.

Limitations

- Increase in CFC-11 emissions 2012–2017 suggests ban on production was broken by one or more countries.

Convention on International Trade in Endangered Species (CITES, 1975)

Agreement between 184 governments to ban the trade in highly vulnerable plant and animal species and the products made from them. Aims to prevent extinction, illegal poaching, and trafficking. High levels of exploitation of species also leads to habitat loss, affecting non-threatened plants and animals.

REVISION TIP

For the Montreal Protocol and CITES, bullet point what they do and why they are important.

Successes

- Species protection and population recovery (e.g. Southern White Rhino).
- Increased public awareness.
- International cooperation.
- Ongoing amendments to include more threatened species.

Limitations

- Difficult to enforce due to weaker border controls in some countries.
- High demand for some illegal products still exists.
- Criminal gangs and corruption make it harder to implement regulations.
- Limited financial resources of countries to enforce rules.
- Not all species are covered by CITES.

11 Migration, identity and sovereignty

11.9 Environmental IGOs

Oceans and rivers

UN Convention on the Law of the Sea (UNCLOS) (1982)

Rules and laws governing use of oceans and their resources. Coastal states have rights to exploit an Exclusive Economic Zone (EEZ) extending up to 200 nautical miles from the coast but must also protect marine ecosystems. All nations have the right to fish on the high seas but have to adopt measures to manage and conserve living resources, and prevent and control marine pollution. They have to cooperate with other nations.

Successes

- Regulations have helped to mitigate pollution in some areas, positively impacting the marine ecosystems.

Limitations

- Overexploitation of resources as coastal states prioritise their own interests.
- Hard to enforce.
- Some species need specific IGOs to focus on their protection, e.g. International Whaling Commission.
- Does not directly address challenges in specific regions, e.g. the Arctic Ocean where global warming is increasing access to resources.

Helsinki Water Convention (1992)

The UN Convention on the Protection and Use of Transboundary Watercourses and International Lakes (Water Convention):

Promotes the sustainable management of shared water resources, prevents conflicts, promotes peace and regional integration, and implements the related Sustainable Development Goals. Transboundary water resources are important as their use by one member state can directly impact another. The impacts of climate change are affecting water security in many countries so management of the allocation and quality of the water resources available is important.

Successes

Cooperation between countries has:

- led to agreements on the allocation and management of resources along specific rivers.
- encouraged scientific research to improve monitoring and improvement of water quality.

Limitations

- Not all countries belong to the convention.
- Lack of resources make it difficult to enforce.
- It is difficult to resolve conflicts as it is often hard to prove overexploitation of water in a particular region, e.g. upstream.
- Rivers can flow through many countries, making a fair balance of rights within the drainage basin difficult.

Monitoring the environment

Millenium Ecosystem Assessment (MEA) (2001)

Used global research to assess the consequences of changing ecosystems for human development. Reported on conditions and trends within the world's ecosystems and assessed the products they provide (water, food, natural resources, flood control).

Main findings: over past 50 years humans have changed ecosystems more rapidly than ever before; ecosystems have been used to meet the rapidly growing demands for natural resources from a growing population; while there have been net gains in human development, many ecosystems have been degraded.

Successes

- Identified key issues surrounding ecosystem degradation.
- Increased global awareness of biodiversity loss.
- Some decision-makers used results to inform environmental management policies.

Limitations

- One-off summary with no legally binding outcomes for countries to adhere to.

REVISION TIP

Create memory maps to explain what UNCLOS, the Helsinki Convention and the MEA are.

Antarctica

Antarctic Treaty System (ATS, 1959)

Gives Antarctica special status. It is to be used for peaceful purposes only, with a focus on scientific research. Military activity is banned. Includes specific environmental agreements covering the protection of marine and terrestrial life and the exploitation of minerals.

56 parties have signed the treaty, including those countries who operate research stations and those who have territorial claims to parts of the Antarctic.

REVISION TIP

Make a one-minute audio recording to explain how the Antarctic Treaty System works.

Successes

- No active conflicts on the continent.
- Nuclear-free zone.
- International scientific cooperation.
- Environmental protection of species.

Limitations

- Not all countries have signed.
- Geopolitical relationships have evolved since, so more countries want a say on how the region is used and managed.
- Will need to respond to increased pressure linked to climate change and resource exploitation, including tourism.

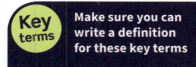 **Make sure you can write a definition for these key terms**

atmosphere   ATS   biosphere   CITES
Helsinki Water Convention   hydrosphere   MEA
Montreal Protocol   UNCLOS

Learn the answers to the questions below, then cover the answers with a piece of paper and write as many as you can. Check and repeat.

Questions	Answers
1 List three purposes of the UN.	Three from: international peace and security / develop friendly relations between countries / solve international economic, social, cultural or humanitarian problems / human rights / coordinate nations to work towards goals
2 What are the two forms of UN intervention?	Economic sanctions and military intervention
3 What is a SAP?	Structural Adjustment Programme: the IMF and World Bank lend money to developing countries to increase economic development but with conditions
4 What is the WTO?	World Trade Organisation: it sets global trade rules and regulations and settles trade disputes
5 What is the HIPC initiative?	Heavily Indebted Poor Countries initiative: countries gain debt relief on loans if they meet set criteria
6 What was NAFTA?	North American Free Trade Agreement: a free trade agreement between USA, Canada and Mexico
7 What is the Montreal Protocol?	Agreement to ban the production and use of CFCs
8 What is CITES?	Agreement concerning plant and animal trading
9 What is UNCLOS?	Agreement concerning the governance of the sea, including its environmental management
10 What is the Helsinki Water Convention?	Agreement to enable countries to manage the supply and quality of water from transboundary rivers and lakes

Put paper here

Previous questions

Now go back and use these questions to check your knowledge of previous topics.

Questions	Answers
1 What is a nation state?	Sovereign territory recognised as an independent state
2 What is homogeneity?	High level of similar culture, language and beliefs within a society
3 What is a tax haven?	Country or territory that offers very low or zero tax rates
4 When did India gain independence?	1947

Put paper here

11 Migration, identity and sovereignty

11.10 National identity

Ways nationalism is reinforced

Nationalism is a complex term but links to people's identification and affinity with and support for their own nation, often at the expense of other countries' interests.

Education	Sport
UK government introduced 'Fundamental British Values' into its education system (2014). Schools should promote democracy, the rule of law, individual liberty, and mutual respect and tolerance of those with different faiths and beliefs.	The national team in particular sports or sporting events can act as a focus of pride in the nation. Supporting the team may involve flags, songs and fan clothing. Sport can join people of different groups within a country, e.g. French-, Flemish- and German-speaking communities in Belgium all support the Belgian national football teams.
Schools in other countries may sing their national anthem regularly or have an image of the country's flag or leader in their classrooms.	It can also highlight differences, e.g. England, Wales, Scotland and Northern Ireland have their own national football teams.

Political parties

Political parties can attempt to gain voters by focusing on nationalism through policies and the symbols they use. The UK Conservative Party has used red, white, and blue logos and currently uses an oak tree as their symbol. Political parties stress the importance of national institutions, e.g. the NHS.

Identity and loyalties

National identity and **national loyalties** have historically been tied to different systems and ideas. These can be based on very generalised concepts, or past eras, and can cause stereotyping. People may feel different degrees of identity and loyalty to different systems. In the UK, the royal family plays an important role in national identity for many. It has evolved into a constitutional monarchy and has a soft-power role, promoting Britain's national identity at home and abroad.

Identity and loyalties have been historically tied to:

Legal systems	People may take pride in their country's legal system if it promotes justice and fairness. In the UK, the role of the jury is a key part of the legal system. People accused of serious offences are tried by 12 members of the public, which some people believe is the best way to reach fair, reasonable, and unbiased verdicts.
Methods of governance	People may be proud of a nation's form of government. They may be loyal to a particular party or leader (e.g. North Korea) or to a particular system (e.g. democracy).
National 'character'	This can refer to the shared beliefs or lifestyles, or perceptions of the personality traits, of people living in a country. Stereotypes may develop over time. Generalised differences can be put forward between people of different countries and of different regions in those countries. These traits can be linked to cultural characteristics, e.g. behaviour, language and dialect, traditions, food. Some nations may be perceived as more nationalist or patriotic than others.

11 Migration, identity and sovereignty

11.10 National identity

The 'English' countryside

- The natural environment of a country can reinforce national identity.
- **The English countryside** has played an important part in legends, literature, music, and art as well as propaganda and advertising.
- Rural village life evokes nostalgia for a bygone preindustrial era with perceptions of stability, calmness and simplicity.
- Images of the English landscape were used in wartime recruitment posters to encourage people to sign up to the military to protect their country.

> **REVISION TIP**
>
> Try to think about what influences the identity of and loyalty to a particular country.

▲ **Figure 1** *Specific natural landmarks such as the White Cliffs of Dover are important symbols of the nation and have been associated with homecoming, history, and defence.*

National identity and globalisation

- The globalised world has increased the multinational and multiethnic composition of different countries. Migration of people and the rise of internet-based communications has led to both the exchange of cultures and the reinforcement of different cultures within a country.
- Diversity can challenge ideas of national identity and loyalties.
- People may question whether they are a global citizen rather than a national of a particular country.
- People may embrace more than one national identity, e.g. by supporting more than one national sports team, or because they are 'third-culture kids' – children brought up by two people of different nationalities (e.g. British and French) while living in another country (e.g. Belgium).
- National identity may change over time because of increased diversity.

> **REVISION TIP**
>
> Make a mind map of the ways in which globalisation makes the issue of national identity more complex.

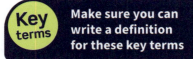

Make sure you can write a definition for these key terms

national character   national identity
national loyalties   the English countryside

11.11 Challenges to national identity

'Made in Britain'?

- Many UK-based companies are owned by foreign companies (which may be owned by foreign states).
- Globalisation has made it easier for TNCs and other overseas investors to buy into or take over companies as well as purchase land and property in other countries.
- Increased foreign ownership has challenged the idea of the British brand and the **'Made in Britain'** label.
- Profits from products made in the UK may be part of a global business.

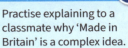

REVISION TIP

Practise explaining to a classmate why 'Made in Britain' is a complex idea.

EDF

- EDF is a British energy company which is owned by Electricitié de France, a French state-owned energy company. Energy is important for the national economy, society, and security.
- Advantages: increased efficiencies and access to new technologies and expertise through overseas ownership.
- Disadvantages: overseas ownership of key infrastructure may cause challenges, e.g. the UK government being unable to control energy prices for UK customers.

Jaguar Land Rover (JLR)

- JLR began as British vehicle manufacturers. Tata Motors Ltd, an Indian TNC, bought Jaguar and Land Rover when both companies had financial issues, keeping some vehicle production in the UK.
- JLR employs thousands of people in the UK and are developing new technologies in the UK. They promote the 'Spirit of Britishness' of their cars through their advertising.
- Cars may be built in the UK but their components may be made in other countries.
- Some people are concerned that TNCs like JLR may, in the future, move production to countries where costs are cheaper, increasing their profits but cutting UK jobs.

Westernisation

- **'Westernisation'** is the promotion of European and North American culture around the world. It is generally dominated by US culture.
- Globalisation has helped spread capitalist ideas and consumer culture in different countries.
- Large retailing and entertainment corporations promote 'western' ideas, values, and behaviours as well as specific goods and services. Following the capitalist model allows them to increase their profits by accessing the global market. Smaller local companies find it difficult to compete, threatening their business.
- Critics suggest that the spread of 'westernisation' weakens national identity as traditional products and customs may be eroded. Others suggest that people actively seek local alternatives, reinforcing national identity.

REVISION TIP

Make a mind map to encapsulate the idea of 'westernisation'.

11 Migration, identity and sovereignty

11.11 Challenges to national identity

Property, land, and business ownership

- Non-national ownership of property, land, and business can challenge national identity and the national economy.
- Adoption of overseas brands and lifestyles can reduce the unique quality of a nation's culture.
- Companies based overseas may apply pressure on national governments, affecting the decisions made. This may influence national identity.

▲ **Figure 1** *London is a global city popular with overseas buyers*

London is a global city popular with overseas buyers. Wealthy people buy property in affluent areas such as Mayfair as an investment.

Increased overseas ownership in some areas may lead to a 'less British' feel in some places. House prices may also increase, making property unaffordable for local people.

Non-national ownership in London

Many London landmarks are owned by investors from overseas. The Qatari Investment Authority (QIA) has invested £386 billion including in Canary Wharf.

Property may be bought by non–nationals who register their assets in tax havens, so paying less tax into the UK economy. Non-nationals may also face sanctions on their investments because of external factors, e.g. after the Russian intervention in Ukraine.

Foreign ownership of TNCs in the UK: China

Many British companies are controlled or partially owned by Chinese companies. The China Investment Corporation (CIC) is involved in businesses in the transport, energy, entertainment, and education sectors.

REVISION TIP

Create flashcards with examples of how ownership of property or TNCs is increasingly non-national.

Key terms Make sure you can write a definition for these key terms

'Made in Britain'   westernisation

11.12 Disunity within nations

Nationalist movements

Some groups within a country may seek **independence** from the nation they live in. Nationalist movements form within regions with particular ethnic, language, or religious beliefs. Their goal is **secession**.

If a region within an EU country gained independence, they would have to apply to rejoin the EU as an independent state, meeting certain conditions. Other EU member states, including the country the state had just gained independence from, would have to agree.

Catalonia

- Catalonia is an autonomous region of northeast Spain. It has a long history of seeking independence.
- Catalans have their own language (Catalan), food and culture.
- Catalan nationalists argue that Catalonia, which is a wealthy region, puts more into the Spanish economy than the region receives.

> **REVISION TIP**
>
> Find a map of a country that has a strong nationalist movement, identify the area with the nationalist movement and add annotations with points to show why they want to become independent.

Demonstrations leading to a non-binding referendum within Catalonia where people voted in support of independence.

Catalonia remains part of Spain, but the nationalist movement still pushes for secession.

1979 > 2013–2014 > 2017 > Post 2017

Post-Franco government recognised Catalonia as an 'autonomous community' but the powers given were not sufficient for some Catalan nationalists

Referendum with 90% of voters voting for Catalan independence. Results not recognised by the Spanish government or the EU.

Political tensions

Some BRICS and emerging countries, and regions within them, have benefited more than others from globalisation, causing:

- Increase in regional income inequality between urban and rural areas
- Increase in new middle classes with increased spending power
- Increase in cost of living
- Some groups may feel exploited by globalisation
- Loss of government trust leading to protests and political divisions

- Rise in separatist and nationalist groups
- Rise in ethnic tensions as some minority groups may feel left behind by globalisation
- Tensions over exploitation of natural resources
- Tensions over human rights issues
- External and internal migration.

> **REVISION TIP**
>
> Remember, you need to know how the costs and benefits of globalisation cause tensions within BRICS nations.

11 Migration, identity and sovereignty

11.12 Disunity within nations

'Fragile states'

A **'fragile state'** is a state where the government is no longer in control. Political and economic systems are weak, and huge disparities exist between the powerful elite (often supported by overseas investors) and the rest of the population. The government may no longer be able to assert authority over national borders.

Inability to maintain law and order

Low levels of economic development

Frequent changes of leadership

Governments may be ineffective due to:

Conflicts between ethnic or religious groups, or civil war

Inability to tackle population, health and environmental issues

Somalia, Northeast Africa

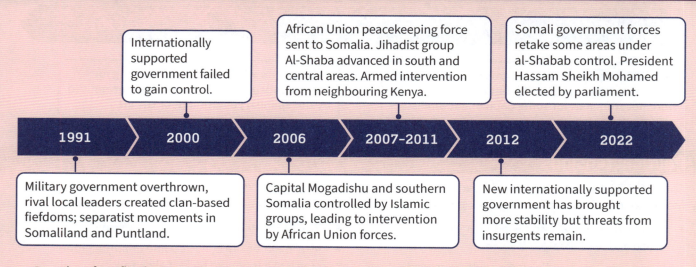

Internationally supported government failed to gain control.

African Union peacekeeping force sent to Somalia. Jihadist group Al-Shaba advanced in south and central areas. Armed intervention from neighbouring Kenya.

Somali government forces retake some areas under al-Shabab control. President Hassam Sheikh Mohamed elected by parliament.

1991 — 2000 — 2006 — 2007–2011 — 2012 — 2022

Military government overthrown, rival local leaders created clan-based fiefdoms; separatist movements in Somaliland and Puntland.

Capital Mogadishu and southern Somalia controlled by Islamic groups, leading to intervention by African Union forces.

New internationally supported government has brought more stability but threats from insurgents remain.

- Decades of conflict has contributed to underdevelopment: low life expectancy, low GDP per capita and low HDI.
- Media is fragmented and dominated by radio, with poor internet and TV coverage.
- Somalia attracts FDI and has been given international aid. Effectiveness is limited by poor infrastructure, corruption and ongoing conflict.

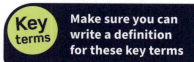 **Key terms**

Make sure you can write a definition for these key terms

'fragile state'   independence   secession

RETRIEVAL

Learn the answers to the questions below, then cover the answers with a piece of paper and write as many as you can. Check and repeat.

Questions | Answers

	Questions		Answers
1	What is nationalism?		People's identification and affinity with, and support for, their own nation, often at the expense of other countries' interests
2	Give one way in which nationalism can be reinforced.		One from: education / sport / political parties
3	What are 'third-culture kids'?		Children brought up by two people of different nationalities whilst living in another country
4	EDF is owned by a company from which country?		France
5	Indian TNC Tata Motors Ltd owns which major car company?		Jaguar Land Rover
6	What is 'westernisation'?		The promotion of European and North American culture around the world
7	What is a 'fragile state'?		A state where the government is no longer in control
8	Give three systems or ideas to which national identity and loyalties may be tied.		Legal systems, methods of governance, 'national character'
9	Give two reasons why governments may be ineffective in a 'fragile state'.		Two from: conflicts between ethnic or religious groups, or civil war / inability to maintain law and order / inability to tackle population, health and environmental issues / frequent changes of leadership / low levels of economic development

Put paper here (repeated along the central divider)

Previous questions

Now go back and use these questions to check your knowledge of previous topics.

Questions | Answers

	Questions		Answers
1	What is the WTO?		World Trade Organisation: it sets global trade rules and regulations and settles trade disputes
2	What are the two forms of UN intervention?		Economic sanctions and military intervention
3	What is a SAP?		Structural Adjustment Programme: the IMF and World Bank lend money to developing countries to increase economic development but with conditions
4	What is the HIPC initiative?		Heavily Indebted Poor Countries initiative: countries gain debt relief on loans if they meet set criteria
5	What was NAFTA?		North American Free Trade Agreement: a free trade agreement between USA, Canada and Mexico

Put paper here (repeated along the central divider)

PRACTICE

Exam-style questions

1   Explain why globalisation has changed the demand for labour.   **(4)**

2   Explain why some countries allow freedom of movement within their borders.   **(4)**

EXAM TIP

For a 4-mark 'Explain' question, your short paragraph answer should provide a reasoned explanation of how or why something occurs, demonstrating your understanding of the topic through your justification and/or examples.

3   Explain why national identity may be tied to a country's legal system.   **(4)**

4   Explain how globalisation has encouraged the growth of tax havens.   **(4)**

5   Study **Figure 1**.

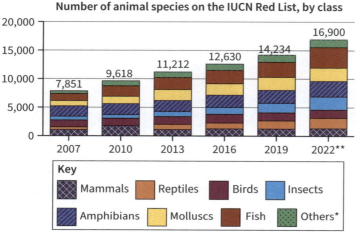

* other invertebrate (spineless) animals, such as crustaceans, corals and arachnids (spiders, scorpions)
** preliminary

▲ **Figure 1** *Threatened species IUCN red list, 2019*

i)   Calculate the percentage change to 1 decimal place, stating whether it is an increase or decrease, between 2007 and 2022 for the total number of animal species on the IUCN Red List. Show your working.   **(3)**

ii)   Suggest reasons for the overall change in the number of animal species on the IUCN Red List in **Figure 1**.   **(3)**

6   Explain why not all nation states are universally recognised.   **(6)**

7   Explain why migration changes the cultural composition of nation states.   **(6)**

EXAM TIP

8-mark 'Explain' questions require a longer paragraph answer.

8   Explain why some migrants move from rural areas to urban areas.   **(8)**

9   Explain why globalisation may cause tension within BRIC countries. **(8)**

EXAM TIP 🎯

These 8-mark questions have all 8 marks for AO1, so explain your knowledge and understanding of the concepts in the question in detail and use place specific examples where possible.

10   Explain why the impact of Structural Adjustment and HIPC policies on the developing world is disputed. **(8)**

11   Explain why the management of oceans by IGOs is challenging. **(8)**

12   Explain how ownership of property and land by foreign nationals can challenge national identity. **(8)**

EXAM TIP 🎯

20-mark 'Evaluate' essay questions require you to measure the value or success of something. Your essay answer should give a balanced and substantiated judgement and needs to review the information, then bring it all together by reaching a conclusion that draws on evidence (strengths, weaknesses, alternatives, data).

13   Evaluate the successes and challenges that the UN has experienced when addressing global issues. **(20)**

14   Evaluate the role of international organisations in the management of the global economy. **(20)**

15   Evaluate the effectiveness of the Antarctic Treaty System in the management of the challenges faced by Antarctica. **(20)**

16   Evaluate the view that national identity is threatened in 'fragile states'. **(20)**

EXAM TIP 🎯

Use examples and place contexts that you have studied to exemplify your points.

12 Fieldwork and skills

12.1 Fieldwork

Preparation for fieldwork

Before fieldwork is conducted, preparation involves:

- background reading into the geographical theories and concepts to ensure that the enquiry is connected to relevant theories and that there is literature to support the analysis of findings

- drawing up aims and **objectives** for the enquiry, including an overall **research question** and sub-questions, which may take the form of a **hypothesis** to be tested

- ensuring that research aims and objectives stem from key geographical concepts and theories that can be tested, and existing geographical literature that can be referenced

- planning the collection of **primary data** and the use of **secondary data**

- planning the **sampling** procedures that will be used for primary data collection to eliminate bias.

> **REVISION TIP**
>
> Ensure that you can justify the use of different sampling techniques, as well as being able to discuss their advantages and drawbacks.

A sampling strategy should be devised to ensure the whole population is accurately represented in the data collection.

Sampling strategies can be: systematic (every 5 m), stratified (representative of the proportions of sub-groups in a data set), random (selection through random number generators), opportunistic (asking passers-by to complete a questionnaire), as well as sampling across a transect, an area, or using point sampling.

A risk assessment should be carried out to identify health and safety issues in the field e.g. working close to the sea.

Steps should be put in place to mitigate the risks e.g. choosing study sites away from the coastline and knowing the tide times.

An OS map can be used to plan primary data collection as the whole sample area can be seen, access can be determined and physical features, such as relief and water sources, can be seen.

Ethical considerations involve:

- Minimising environmental damage
- Ensuring anonymity and confidentiality when using participants
- Ensuring consent is obtained for data collection at the field site
- Ensuring that no offence is caused when interacting with participants.

Primary field methodologies

Primary data can be **quantitative** (numerical) or **qualitative** (words). The methods used to collect the primary data should be appropriate for the aims of the investigation. Secondary data can be used to help reach the aims of geographical enquiry, e.g. house prices, crime data, weather data, census data.

> **REVISION TIP**
>
> Ensure that you know the advantages and disadvantages of using different primary data collection techniques.

- Environmental or residential quality survey
- Surveys, questionnaires and interviews
- Beach profile and sediment survey
- Vehicle or pedestrian count

Primary data collection techniques

- Dune profile and vegetation analysis
- Clone town and land use survey
- Participant observation

Analysing field data and representing results

Once data has been collected it is processed and presented using relevant graphical and cartographical techniques.

- Graphical and cartographical techniques are used to present quantitative data and analyse patterns, trends and relationships.

- **Geo-located** data is data which is presented on a map, such as a proportional symbols map displaying vehicle or footfall counts on a base map of a central urban area.

- GIS can be used to present data as different types of data collected can be layered on a base map to reveal and better analyse patterns.

- Other techniques such as annotated photographs, field sketches, quotes and word clouds can be used to present qualitative data.

> **REVISION TIP**
>
> Ensure that you can choose appropriate presentation techniques by understanding the effectiveness (strengths and weaknesses) of each one.

> **REVISION TIP**
>
> Look at 12.2 to see examples of graphs and maps that can be used for data presentation.

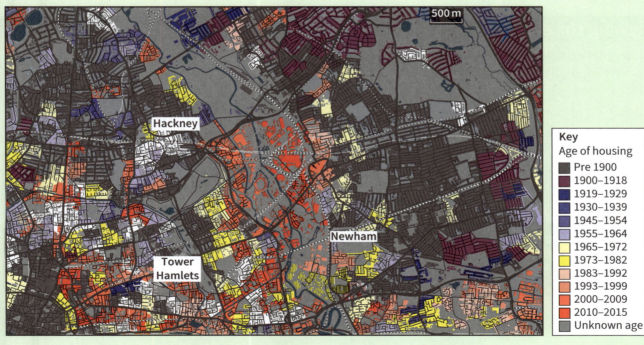

▲ **Figure 1** An example of GIS data plotted on a base map; the age of housing mapped across the Queen Elizabeth Olympic Park and surrounding areas in East London

 **KNOWLEDGE**

12 Fieldwork and skills

12.1 Fieldwork

Analysing data

- Quantitative data can be analysed using **statistical** techniques such as:
  - measures of central tendency (mean, median, mode)
  - measures of dispersion (range, inter-quartile range, standard deviation)
  - inferential and statistical techniques (Spearman's rank and chi-squared).
- Analysis of primary and secondary data should involve interpreting the patterns and trends shown on the presented data.
- Different sets of primary and secondary data can be linked together and compared.
- Analysis should link to background reading on the geographical theories and concepts that are relevant to the enquiry.
- Analysis should be structured around the sub-questions.

 REVISION TIP

Ensure you can calculate all of the statistical techniques and can interpret the outcomes of statistical techniques such as Spearman's rank correlation coefficient. Look at the Skills section to remind yourself.

Drawing conclusions

A summary of the key findings of the investigation should be made.

Drawing conclusions

Essentially, sub-questions should be answered and comments made as to whether the hypothesis has been proved or disproved.

Conclusions should relate back to the original aims and objectives of the enquiry.

Reviewing the enquiry

- All stages of a fieldwork-based enquiry should be critically **evaluated**, e.g. were appropriate methods chosen? Were appropriate sampling methods used?
- Suggestions of appropriate improvements should be made.
- Limitations of the enquiry should be addressed and comments on validity, reliability and accuracy can be made.
- Consideration should also be given as to how the enquiry could be further developed.

 REVISION TIP

Ensure that you can evaluate the planned data collection methods of an unfamiliar fieldwork-based enquiry.

 Key terms
Make sure you can write a definition for these key terms

evaluate   geo-located   hypothesis   objective
primary data   qualitative   quantitative
research question   sampling   secondary data   statistical

RETRIEVAL

Learn the answers to the questions below, then cover the answers column with a piece of paper and write as many as you can. Check and repeat.

Questions | Answers

#	Question	Answer
1	Why is background reading important in preparing for fieldwork?	To ensure that the enquiry is connected to existing theories and that there is literature to support the analysis of findings
2	Why do sampling techniques need to be planned?	To eliminate bias in primary data collection
3	What is quantitative data?	Numerical data
4	What is qualitative data?	Non-numerical data e.g. words
5	Give two examples of secondary data.	Two from: House prices / crime data / weather data / census data
6	Give two examples of primary data collection techniques for physical geography enquiry.	Two from: beach profile / sediment survey / dune profile / vegetation analysis
7	What is geo-located data?	Data which is presented on a map
8	How can qualitative data be presented?	Annotated photographs, field sketches, quotes and word clouds
9	What are the three measures of central tendency?	Mean, median, mode
10	What are three measures of dispersion?	Range, inter-quartile range
11	What is Spearman's rank?	A statistical technique used to discover the strength of a link between two sets of data
12	What should conclusions of a geographical enquiry relate back to?	The original aims and objectives of the enquiry

Put paper here

12 Fieldwork and skills

12.2 Representations of geographical information

Dot maps

- **Dot maps** use small dots that each represent the same value (e.g. 7,500 people) to show the distribution of data within an area.

- The **spatial variation** of values across the map area is shown by interpreting **clusters** and absences of dots.

- The absence of a dot may not always imply that there is no data there – the threshold for a dot (e.g. 7,500 people) may not have been reached.

- A cluster of small dots may make it difficult to see total values represented.

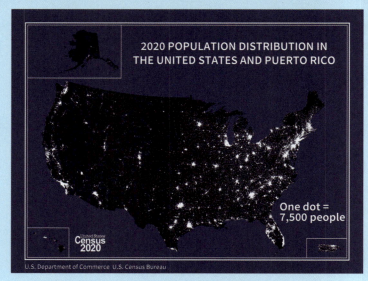

▲ *Figure 1* USA population distribution, 2020

Kite diagrams

- **Kite diagrams** are often used to show distribution of plant species across an area.

- Data is usually collected along a **transect** (e.g. distance from the shore).

- The data is plotted in two halves, half above the line and half below.

- Interpreting the width of the kite created by plotting the points shows the amount of that item present (e.g. species abundance).

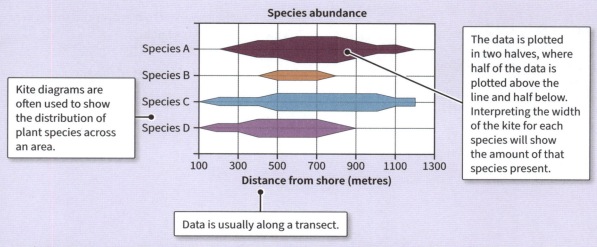

▲ *Figure 2* A kite diagram showing the spatial variation of species across a transect

Scatter graphs

- **Scatter graphs** show the relationship or correlation between two variables.
- A line of best fit can be drawn and **anomalies** identified to aid analysis.
- A **line of best-fit** does not have to pass through the origin.

REVISION TIP

Look at an example of a scatter graph on page 15.

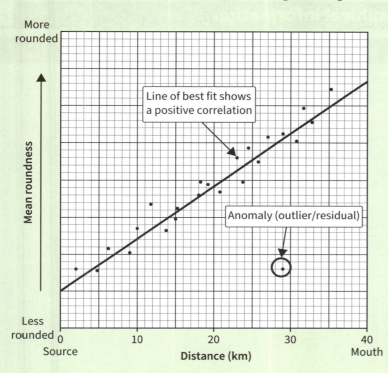

Line of best fit shows a positive correlation

Anomaly (outlier/residual)

◀ **Figure 3** *This scatter graph demonstrates the positive relationship between river particle roundness with distance from the source*

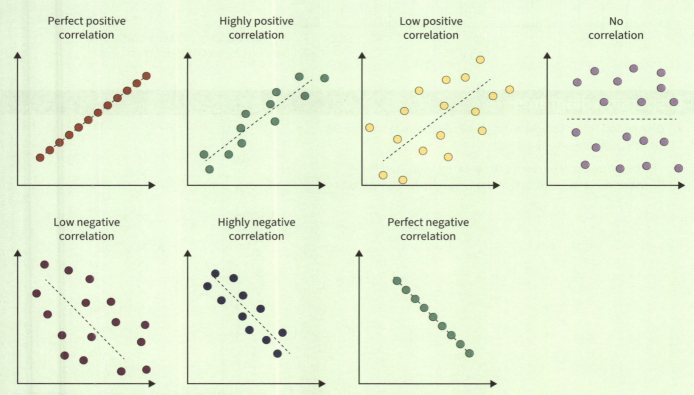

Perfect positive correlation

Highly positive correlation

Low positive correlation

No correlation

Low negative correlation

Highly negative correlation

Perfect negative correlation

▲ **Figure 4** *How close the plotted data sits to the line of best-fit shows how strong the relationship between the two variables is*

12 Fieldwork and skills

12.2 Representations of geographical information

Logarithmic scales

- Graphs can use **logarithmic scales** when the data to be presented has a large range.
- The scale increases by a power of 10 for each interval, i.e. 10 to 100, 100 to 1000, 1000 to 10,000.
- Greater space is given to smaller values, whilst less space is given to larger values.

◄ **Figure 5** *Semi-log scales have one linear axis and one logarithmic axis*

Dispersion diagrams

- **Dispersion** refers to how a data set is distributed and can be shown on a dispersion diagram.
- The range and any outliers can be easily seen, and the distribution of more than two data sets can be compared.

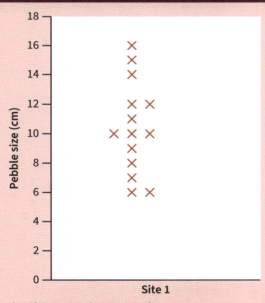

▲ **Figure 6** *A dispersion diagram*

Geographical images

Photographs contain a lot of useful geographical data so can be used for annotation and interpretation.

- Aerial photos are those which show the landscape from above and are usually taken from an aircraft.

▲ **Figure 7** *An aerial photograph of London*

- Ground photography images are much smaller scale and show features more closely.

▲ **Figure 8** *A photo of Malham Cove taken at ground level*

- Satellite images are created using data from satellites in space and show much wider areas with less visible detail.

REVISION TIP

Look at other ground, aerial and satellite photographs on pages 36, 52 and 196. What information can you glean from each one?

◀ **Figure 9** *This satellite image can be used to see the features and extent of the Ganges delta in India and Bangladesh*

Geographical Information Systems (GIS)

- **GIS** involves layering geographical data on base maps on a computer.
- It makes data much easier to analyse; patterns and relationships can be identified as they can be visualised on the map.
- The scales, size and detail on maps can easily be changed, and complex data can be processed quickly.

REVISION TIP

Look at an example of GIS data plotted on a base map on page 313.

 Key terms Make sure you can write a definition for these key terms

anomaly   dispersion   dot map   GIS   kite diagram
line of best-fit   logarithmic scales   scatter graph
spatial variation of values   transect

12 Fieldwork and skills

12.3 Statistical skills

Measures of central tendency

Measures of central tendency are used to describe a data set:

- The mean is used to show the average of a data set.
- The median is used to show the mid-point of a data set.

- The mode is used to show the most common value in a data set.
- The mean is very commonly used but can be influenced by extreme **outliers** and does not give any information about how the data is spread around the mean.

Measures of dispersion

Measures of dispersion are used to show how data is distributed.

- The range is calculated by looking at the difference between the highest and lowest values in a data set.
- The **inter-quartile range** omits extreme outliers and gives the range of the middle 50% of the data in the data set. To obtain it, the mid-point of the bottom 25% of the data (Q1) is subtracted from the mid-point of the top 25% of the data (Q3) to find the inter-quartile range.
- The **standard deviation** can be calculated to show how far a data set is spread around the mean.

> **REVISION TIP**
>
> Make sure you can calculate and interpret a value for the standard deviation.

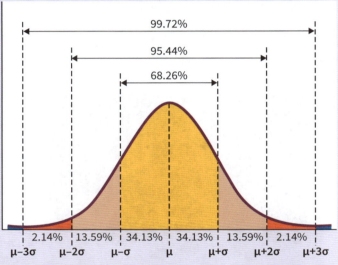

▲ **Figure 10** The normal distribution curve tells us that 68.26% of the data within a data set will lie within 1 standard deviation (σ) either side of the mean.

Gini coefficient and Lorenz curve

- The **Gini coefficient** measures income inequality.
- 0 indicates complete equality or an equal distribution of wealth, and 1 indicates complete inequality where wealth is concentrated with one person.
- Knowing the Gini coefficient value of a country gives a better view of its level of socio-economic development.
- The Lorenz curve is a graphical representation of inequality.

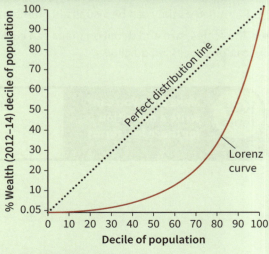

▲ **Figure 11** The Lorenz curve

Inferential and relational statistical techniques

- These statistical techniques are used to make inferences about a data set as a sample of a larger population.

- **t-tests** are used when the sample size is low (less than 30) to see if there is a significant difference between two data sets. It can also be carried out for data which is in pairs.

- **Spearman's rank** correlation gives a value, or correlation coefficient (R_s), to the relationship between two variables. An R_s value closer to −1 indicates a negative relationship and an R_s value closer to 1 indicates a positive relationship. The closer the value is to −1 or 1 the greater the strength of the relationship.

- The **chi-squared test** can be used to see if there is a significant difference between two data sets.

> **REVISION TIP**
>
> Make sure you can calculate and interpret a value for t-tests, Spearman's rank correlation and Chi-squared.

Application of significance tests

- The results of t-tests, Spearman's rank and chi-squared tests can be compared against a **critical values table** to examine how likely it is that the observed result has occurred by chance.

- If the calculated value is higher than the number given in the critical value table at both the 95% and 99% significance level, then it can be said that the result is highly significant.

- For Spearman's rank, the higher the value of R_s, the more significant the result is. The larger the value of chi-squared, the smaller the possibility that there is no significant difference between the two data sets.

> N is the number of values in the data set.

> The 99% or 0.01 significance level is a 1 in 100 probability that the result has occurred by chance.

N	95% (0.05)	99% (0.01)
5	0.90	1.00
6	0.83	0.94
7	0.71	0.89
8	0.64	0.83
9	0.60	0.78

> If the calculated value is higher than the number given in the critical value table at both the 95% and 99% significance level, this indicates that the result is highly significant.

> **REVISION TIP**
>
> When testing the chi-squared value for significance, remember to use N−1 in the critical values table.

▲ *Figure 12 The critical values for Spearman's rank for a data set of between 5 and 9 values*

Key terms Make sure you can write a definition for these key terms

chi-squared test   critical values table   Gini coefficient
inter-quartile range   measures of central tendency
measures of dispersion   Spearman's rank
standard deviation   t-tests

Learn the answers to the questions below, then cover the answers with a piece of paper and write as many as you can. Check and repeat.

Questions

	Answers
1 On a dot map, the absence of a dot implies there is no data for that area. True or false?	False. It may be that the threshold for a dot may not have been reached
2 In what sort of data representation is data plotted half above the line and half below?	Kite diagram
3 Kite diagrams are used to show distribution of an item across an area. True or false?	True. They are often used to show distribution of plant species across an area
4 What is a line of best-fit?	A line drawn on a scatter graph at the point closest to most points marked
5 What does a scatter graph show?	The relationship or correlation between two variables
6 What is an anomaly on a graph?	A point marked on a graph that deviates from the trend shown by the other points and is not what is expected from the data
7 If all plotted data on a scatter graph sits close to the line of best-fit, what does this show?	A strong relationship between the two variables
8 How does the scale on a logarithmic graph increase?	In powers of 10, i.e. 10 to 100, 100 to 1000, 1000 to 10,000
9 What does dispersion mean?	How a data set is distributed
10 What is an aerial photograph?	A photo which shows the landscape from above; it is usually taken from an aircraft
11 What is GIS?	Geographical Information Systems (GIS): layering geographical data on to base maps on a computer
12 What are three examples of measures of central tendency?	Mean, median and mode
13 What is the range?	The difference between the highest and lowest values in a data set
14 Which measure of dispersion gives the range of the middle 50% of the data set?	Inter-quartile range
15 What does the Gini coefficient measure?	Income inequality
16 When is a chi-squared test used?	To see if there is a significant difference between two data sets

Put paper here

OS MAPS SYMBOLS

Symbols on Ordnance Survey maps (1:50 000 and 1:25 000)

ROADS AND PATHS

M I or A 6(M)	Motorway
A 35	Dual carriageway
A 31(T) or A 35	Trunk or main road
B 3074	Secondary road
	Narrow road with passing places
	Road under construction
	Road generally more than 4 m wide
	Road generally less than 4 m wide
	Other road, drive or track, fenced and unfenced
	Gradient: steeper than 1 in 5; 1 in 7 to 1 in 5
Ferry	Ferry: Ferry P – passenger only
	Path

PUBLIC RIGHTS OF WAY

(Not applicable to Scotland)

1:25 000	1:50 000	
- - - - -	....	Footpath
– – – – –	–.–.–.–.–	Road used as a public footpath
+++++	– – – – –	Bridleway
-+-+-+-	-+-+-+-	Byway open to all traffic

RAILWAYS

	Multiple track
	Single track
	Narrow gauge/Light rapid transit system
	Road over; road under; level crossing
	Cutting; tunnel; embankment
	Station, open to passengers; siding

BOUNDARIES

+ – + – +	National
+ + + +	District
–.–..–.–	County, Unitary Authority, Metropolitan District or London Borough
	National Park

HEIGHTS/ROCK FEATURES

50	Contour lines
· 144	Spot height to the nearest metre above sea level

outcrop   cliff   scree

ABBREVIATIONS

P	Post office	PC	Public convenience (rural areas)
PH	Public house	TH	Town Hall, Guildhall or equivalent
MS	Milestone	Sch	School
MP	Milepost	Coll	College
CH	Clubhouse	Mus	Museum
CG	Coastguard	Cemy	Cemetery
Fm	Farm		

ANTIQUITIES

VILLA Roman	× Battlefield (with date)
Castle Non-Roman	* Tumulus/tumuli (mound over burial place)

LAND FEATURES

ruin	Buildings
	Public building
	Bus or coach station
Place of worship	with tower / with spire, minaret or dome / without such additions
°	Chimney or tower
	Glass structure
Ⓗ	Heliport
△	Triangulation pillar
	Mast
	Wind pump; wind generator
	Windmill
+	Graticule intersection
	Cutting; embankment
	Quarry
	Spoil heap, refuse tip or dump
	Coniferous wood
	Non-coniferous wood
	Mixed wood
	Orchard
	Park or ornamental ground
	Forestry Commission access land
	National Trust – always open
	National Trust, limited access, observe local signs
	National Trust for Scotland

TOURIST INFORMATION

P	Parking
P&R	Park & Ride
V	Visitor centre
i	Information centre
☎	Telephone
	Camp site/ Caravan site
	Golf course or links
	Viewpoint
PC	Public convenience
✗	Picnic site
	Pub/s
	Museum
	Castle/fort
	Building of historic interest
	Steam railway
	English Heritage
	Garden
	Nature reserve
	Water activities
	Fishing
☆	Other tourist feature
	Moorings (free)
	Electric boat charging point
	Recreation/leisure/ sports centre

WATER FEATURES

Marsh or salting   Towpath Lock   Slopes   Cliff   High water mark
Aqueduct   Canal   Ford   Flat rock   Low water mark   Lighthouse (in use)
Lake   Weir   Normal tidal limit   Sand   Dunes   Lighthouse (disused)   Beacon
Footbridge   Bridge   Mud   Shingle
========= Canal (dry)

OXFORD
UNIVERSITY PRESS

Great Clarendon Street, Oxford, OX2 6DP, United Kingdom

Oxford University Press is a department of the University of Oxford. It furthers the University's objective of excellence in research, scholarship, and education by publishing worldwide. Oxford is a registered trade mark of Oxford University Press in the UK and in certain other countries.

Written by David Alcock, Kyle McFarlane, Rebecca Priest, Paul Schofield, Lucy Scovell and Nadine Tunstall

The moral rights of the author have been asserted

First published in 2024

British Library Cataloguing in Publication Data

Data available

978-1-382-05244-3

978-1-3-820-5245-0 (eBook)

10 9 8 7 6 5 4 3 2 1

The manufacturing process conforms to the environmental regulations of the country of origin.

Printed in the UK by Bell and Bain Ltd, Glasgow.

Acknowledgements
The publisher and authors would like to thank the following for permission to use photographs and other copyright material:

Photos: p33: Milankovitć-Cycles: Hannes Grobe, Alfred Wegener Institute for Polar and Marine Research; **p36:** Fotimageon / Shutterstock; **p50:** Leonard Zhukovsky / Shutterstock; **p52:** Barry Vincent / Alamy Stock Photo; **p57:** Maridav / Shuttertock; **p60(t):** Drepicter / Alamy Stock Photo; **p60(b):** ARCTIC IMAGES / Alamy Stock Photo; **p61:** All Canada Photos / Alamy Stock Photo; **p64:** Matthijs Wetterauw / Shutterstock; **p66:** Becky Stares / Shutterstock; **p81:** Matthew J Thomas / Shutterstock; **p83(l):** Julian Cartwright / Alamy Stock Photo; **p83(r):** ZUMA Press, Inc. / Alamy Stock Photo; **p85(t):** M. Palmer / Alamy Stock Photo; **p85(b):** Michael Hawkridge / Alamy Stock Photo; **p86(t):** atthle / Shutterstock; **p86(m):** Fulcanelli / Shutterstock; **p86(b):** James Osmond / Alamy Stock Photo; **p87(t):** David Robertson / Alamy Stock Photo; **p87(b):** Rob Read / Alamy Stock Photo; **p92:** Stanley Dullea / Shutterstock; **p138:** Pecold / Shutterstock; **p142, 143:** Reach Plc / Evening Gazette; **p171:** David Alcock; **p196(t):** Avalon.red / Alamy Stock Photo; **p196(b):** NASA; **p199(a):** szefei / Shutterstock; **p199(b):** Pakhnyushchy / Shutterstock; **p199(c):** Daniel J. Rao / Shutterstock; **p199(d):** Netta Arobas / Shutterstock; **p199(e):** Matt Gibson / Shutterstock; **p202:** Alexander Mazurkevich / Shutterstock; **p224:** Thanakkan Ramesh; **p237:** steve estvanik / Shutterstock; **p242:** Richard Whitcombe / Shutterstock; **p278:** Statista, Source: Amnesty International; **p295:** blurAZ / Shutterstock; **p304:** Adrian Reynolds / Shutterstock; **p306, 319(tl):** Songquan Deng / Shutterstock; **p312:** blickwinkel / Alamy Stock Photo; **p313:** © Consumer Data Research Centre 2016 / © Crown Copyright & Database Right 2014-5; **p316:** US Dept. of Commerce, US Census Bureau, 2020; **p319(tr):** Andy J Billington / Shutterstock; **p319(b):** Universal Images Group North America LLC / Alamy Stock Photo.

Artwork by Aptara Inc., Angeles Peinador, Barking Dog Art, Dave Russell Illustration, Ian West, Kamae Design, Lovell Johns, Mike Connor, Q2A Media, Simon Tegg, and Oxford University Press.

Ordnance Survey (OS) is the national mapping agency for Great Britain, and a world-leading geospatial data and technology organisation. As a reliable partner to government, business and citizens across Britain and the world, OS helps its customers in virtually all sectors improve quality of life.

Text: p27: Adapted from: How Much Economic Damage do Large Earthquakes Cause? The Oklahoma Economist, March 04, 2016. Federal Reserve Bank of Kansas City https://www.kansascityfed.org/oklahomacity/oklahoma-economist/2016q1-economic-damage-large-earthquakes/ Data Sources: NOAA, US Census, UNdata, Eurostat, Japan Statistics Bureau; **p30:** Based on data from: Worlddata.info Earthquakes in Turkey. https://www.worlddata.info/asia/turkey/earthquakes.php WorldData.info, retrieved on 21/04/2024.

Every effort has been made to contact copyright holders of material reproduced in this book. Any omissions will be rectified in subsequent printings if notice is given to the publisher.

The publisher would also like to thank Adam Robbins and Bob Digby for sharing their expertise and feedback in the development of this resource.